PROGRESS IN ANALYTICAL CHEMISTRY

VOLUME 8

PROGRESS IN ANALYTICAL CHEMISTRY

Based upon the Eastern Analytical Symposia

Series Editors:

Ivor L. Simmons   *M&T Chemicals, Inc., Rahway, New Jersey*

and Galen W. Ewing   *Seton Hall University, South Orange, New Jersey*

A Continuation Order Plan is available for this series. A continuation order will bring delivery of each new volume immediately upon publication. Volumes are billed only upon actual shipment. For further information please contact the publisher.

PROGRESS IN ANALYTICAL CHEMISTRY

VOLUME 8

Edited by
Ivor L. Simmons

M&T Chemicals, Inc.
Rahway, New Jersey

and

Galen W. Ewing

Seton Hall University
South Orange, New Jersey

PLENUM PRESS • NEW YORK AND LONDON

Library of Congress Cataloging in Publication Data

Eastern Analytical Symposium, New York, 1975.
  Progress in analytical chemistry.

  (Progress in analytical chemistry; v. 8)
  Includes bibliographical references and index.
  1. Chemistry, Analytic—Congresses. I. Simmons, Ivor L. II. Ewing, Galen Wood,
1914-   III. Title.
QD71.E17 1975   543   76-14896

ISBN 978-1-4684-3326-5   ISBN 978-1-4684-3324-1 (eBook)
DOI 10.1007/978-1-4684-3324-1

Proceedings of the Eastern Analytical Symposium
held in New York, October, 1975

© 1976 Plenum Press, New York
Softcover reprint of the hardcover 1st edition 1976
A Division of Plenum Publishing Corporation
227 West 17th Street, New York, N.Y. 10011

PREFACE

Volume 8 in the series *Progress in Analytical Chemistry* presents a selection of the papers given at the 1975 Eastern Analytical Symposium.

The analytical chemist is under constant pressure not only from the research chemist whose samples he must characterize and control, but also from an ever-increasing group of governmental agencies stimulated by public concern over health and environmental problems, to determine the most sophisticated kinds of compounds as lower and lower levels.

The subjects covered in these papers are wide-ranging, from the analysis of incinerator effluents to the determination of drugs in blood, but through them runs a common theme, the application of the latest instrumental techniques to the problems of analysis.

The authors show how successful they have been in rising to the analytical challenges presented by an increasingly complex world. The editors take this opportunity to thank them for their efforts in producing such excellent papers for publication in so short a time. Our special appreciation goes to Dr. M. W. Miller, who acted as program chairman, and his team of session chairmen: P. R. Brown, L. J. Cline Love, C. Horvath, J. R. Lindsay, and T. C. Rains.

Ivor L. Simmons
Galen W. Ewing

CONTENTS

ANALYSIS OF METALS AND ALLOYS FOR MAJOR CONSTITUENTS AND TRACE ELEMENTS BY ATOMIC ABSORPTION SPECTROSCOPY

J. Y. Marks, G. G. Welcher and R. J. Spellman

Pratt and Whitney Aircraft

Materials Engineering and Research Laboratory
East Hartford, Connecticut   06108

Atomic absorption spectroscopy has proven to be one of the most versatile measurement tools available to the analytical chemist for the characterization of metals and complex alloys for both major constituents and trace elements. Modern instrumentation with digital electronic readout systems and stable, intense light sources has improved the precision of measurements significantly. Sensitivity for most elements used in metallurgical applications is adequate for a precise direct determination; however, for some of the refractory elements preconcentration may be necessary. Most metals and alloys are readily dissolved in common reagent acids or acid mixtures. Standards are then conveniently prepared from stock solutions of the pure elements. Standards should be carefully matched with samples with respect to both cation and anion concentrations to reduce the possibility of interference effects.

MATRIX INTERFERENCE EFFECTS IN MAJOR ELEMENT ANALYSIS

Interferences arising from differences in composition of the matrix between samples and standards are the most serious problems in obtaining accurate analysis for most alloying elements in metallurgical materials. The time necessary for complete vaporization of desolvated salt particles in the flame is a complex function of the composition and size of the salt particle. Only when the standard additions technique is used does the composition of the matrix remain fixed between standards and samples.

1

This type of interference can be particularly difficult
to identify because the enhancement or depression effect
may be linear with the analyte concentration resulting
in linear calibration relationships passing through the
origin. The problem may be reduced, but not eliminated,
by using the high temperature nitrous oxide-acetylene
flame. In addition, relatively large amounts of salt
may be added to the analyte solution to act as a matrix
diluent (1). Potassium chloride is recommended for this
purpose because ionization effects are controlled simultaneously. Applications of atomic absorption to
the determination of specific elements in metallurgical
systems have been reviewed extensively in a recent
treatise (2).

TRACE ELEMENT ANALYSIS

Perhaps the greatest contribution by atomic absorption
spectroscopy to the analytical chemist has been in the
area of trace element analyses. Many metallurgical systems
show detrimental effects of trace elements at parts-per-
billion concentrations. Present Pratt and Whitney Aircraft specifications for alloys used in high stress,
rotating turbine parts allow no more than 0.3 ppm bismuth,
3 ppm selenium or 10 ppm lead. The Society of Automotive
Engineers, Inc., has recently issued an Aerospace Material Specification 2280 for trace element control in
nickel alloy castings. In certain specification classes
maximum allowable amounts of tellurium and thallium are
specified at 0.5 ppm and 5 ppm respectively, and maximum
concentrations of 50 ppm are allowed for antimony, arsenic,
cadmium, gallium, germanium, gold, indium, mercury,
potassium, silver, sodium, thorium, tin, uranium and zinc.
Before the advent of atomic absorption techniques the
determination of many of these trace elements required
the use of several measurement techniques and tedious
separation and concentration steps.

DIRECT FLAME TRACE ELEMENT ATOMIZATION

Sensitivity is adequate for the direct determination of
many trace elements in metals and alloys without separation. The sample is dissolved in an appropriate acid or acid

mixture. After sample dissolution the acid concentration is adjusted to maintain a constant matrix. Standards are prepared by doping solutions prepared to simulate the metal or alloy from stock solutions of constituent elements. Matrix vaporization interference problems are generally not as important in obtaining accurate analyses as in major element determinations because of the relative independence of matrix particle volatility on trace element concentration. The major interference by the matrix material in the direct determination of trace elements using flame atomization is non-analyte absorbance signals arising from scattering, molecular absorbance or atomic absorption by the matrix (3). This interference becomes most serious as the net signal to background ratio is decreased. Losses due to scattering are the most generally encountered source of background light attenuation in trace element analysis by flame atomic absorption in high salt matrices. These losses are most serious at lower wavelengths and follow an approximate λ-4 dependence. Significant molecular absorbance by matrix constituents in the flame is rare in the analysis of metallurgical systems in the spectral region from 2000Å to 3500Å; however, an iron oxide (FeO) absorbance maximum has been identified at 242 nm in the air-acetylene flame when aspirating salts of iron. Several methods have been used to correct for background light attenuation effects in atomic absorption spectroscopy including subtraction of the absorbance at a non-absorbing line near the analytical wavelength, subtraction of the absorbance of a continuum light source at the analytical wavelength or the subtraction of the absorbance of a sample blank, carefully prepared to simulate the composition of the sample with respect to both cation and anion concentration.

Corrections by use of non-absorbing lines near the analytical wavelength are inherently in error because of the variation of scattering, molecular absorbance and matrix atomic absorption with wavelength. Corrections by use of a continuum light source can be in error if there is sharp molecular absorbance structure or matrix atomic absorption lines within the bandwidth of the monochromator. The most accurate method of background correction in flame atomic absorption is the preparation of a blank closely matching the sample conposition.

ELECTRICALLY HEATED FURNACE ATOMIZATION

For those instances where sensitivity is not adequate for
a direct trace element determination by the flame technique,
the graphite furnace atomizer offers an attractive alter-
native. The improved sensitivity of the furnace atomizer
allows direct determination of most important trace
elements in metals and alloys to concentrations of 1 ppm
or lower. Conditions have been reported for the direct
determination of lead, bismuth, selenium, tellurium,
thallium, tin, antimony, gallium, indium and arsenic in
nickel-base alloys (4,5). After acid dissolution, the
sample is diluted to a volume corresponding to 1g metal
per 50 ml. Aliquots of this solution are then transferred to the furnace and atomized according to pre-
determined conditions. Standards are prepared by doping
solutions containing alloy similar to the sample. Alter-
nately the alloy composition may be simulated by combining
the proper proportions of pure element stock solutions
and adjusting acid concentration. A proper understanding
of the role of several variables is important in obtaining
precise and accurate results with the furnace atomization
method.

The acid media chosen for sample dissolution is important
for two reasons. First, many important trace element compounds are volatile and may be lost during the dissolution
of the sample or during pre-atomization heating steps.
For example, arsenic cannot be determined after dissolution
of the sample in an acid mixture containing hydrofluoric
acid. After dissolution, hydrofluoric acid may be added
to the solution without experiencing losses, therefore,
the loss must occur during the dissolution step. Losses
of lead occur in the presence of hydrochloric acid due to
volatilization during preatomization heating of lead
chloride. Secondly, the proper choice of dissolution acid
can play a large role in reducing background absorbance
due to the formation of matrix compounds more volatile
than the analyte. During preatomization heating cycles a
portion of the matrix may be removed resulting in lower
background attenuation during atomization. The acid that
has been found most successful in our laboratory for the
analysis of a wide variety of high temperature nickel-
and iron-base alloys is a mixture of nitric acid and
hydrofluoric acid. An equal volume mixture of each of the

acids and water will rapidly dissolve most metals and alloys. After dissolution is complete the volume is reduced by evaporation to establish a relatively constant acid matrix in the final solution.

Proper choice of instrumentation is important in obtaining the highest sensitivities possible for many trace elements in complex alloys. The accurate control of specifications for bismuth and tellurium at maximum allowable concentrations of 0.5 ppm require optimum utilization of all parameters, including instrumentation. Not all atomic absorption instruments marketed today are capable of achieving the necessary detection limits. Simultaneous background correction is a necessity.

A very important factor in obtaining optimum sensitivity and reproducibility in the determination of trace elements is the proper selection of preatomization and atomization heating parameters. Most manufacturers of electrically heated atomization devices offer at least three different stages of heating; a dry cycle, a char cycle and an atomize cycle. Some manufacturers also offer ramp heating conditions where the temperature is increased at a given rate with time, rather than stepwise to effect a separation of analyte from non-analyte absorbance signals with time. The dry cycle is relatively unimportant in determining sensitivity and precision. Parameters should be chosen to dry the sample in a minimum amount of time with no splattering. The intermediate heating cycle can be quite important in determining sensitivity. Optimum "char" temperatures are quite variable for different trace elements. The maximum allowable char temperature is many times fixed by the boiling points of the analyte species in the atomizer. This is illustrated in Figure 1 for lead by the rapid decrease in sensitivity above 1000°C. The boiling point of lead fluoride is reported as 1290°C. The behavior of thallium is harder to understand. The sensitivity first increases markedly with increasing char temperature, reaches a maximum and then rapidly decreases at temperatures in excess of 650°C. The boiling point of thallium fluoride is 655°C. Background absorbance is strongly dependent on choice of char temperature when the matrix is partially volatilized during this cycle. High background signals are to be avoided because they limit the amount of sample which may be placed in the furnace.

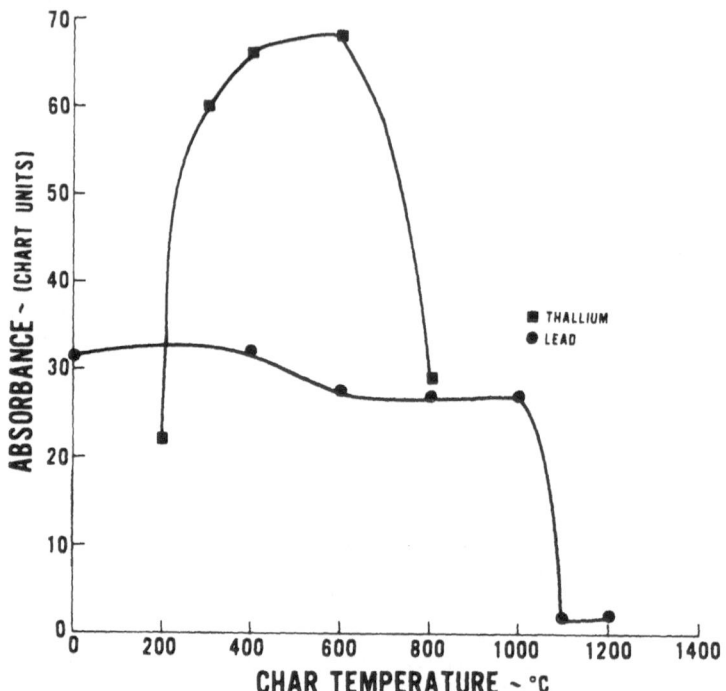

Figure 1.   Effect of varying char temperature on
            sensitivity for lead and thallium.

In addition, inherent errors in the background correction
system may give rise to errors in analytical results if
the errors are not consistent between samples and standards.
The processes occuring during the char cycle are complex
and may include breakdown of thermally unstable species,
reduction of compounds by carbon and formation of stable
carbide species.   Char conditions should be optimized in
the analysis of metals and alloys at the highest temperature possible without introducing imprecision due to
losses of analyte.

The proper choice of atomization temperature is also important in determining sensitivity and reproducibility.
Figure 2 shows systematic studies of sensitivity vs.
atomization temperature for the determination of arsenic
and indium in a Ni-Co-Cr-Al-Ti alloy.   The two curves

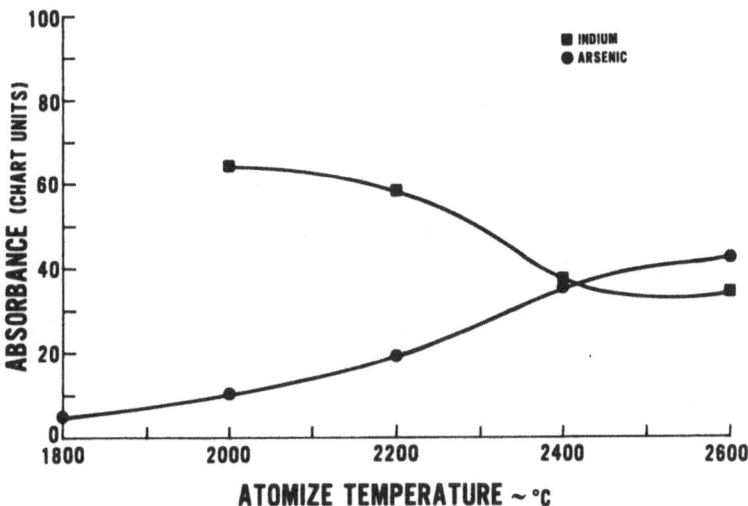

Figure 2.   Effect of varying atomization temperature
on sensitivity for indium and arsenic.

illustrate two extremes in behavior of sensitivity with
atomization temperature. Sensitivity for arsenic increases
steadily with atomization temperature to 2600°C. This is
typical of refractory elements, or those elements which
form stable compounds in the furnace. In the case of
indium the sensitivity first rises with increasing atomization temperature and then apparently decreases with
further increases in atomization temperature. The
initial increase in sensitivity is due to the more favorable conditions for concentration of the indium in the
gas phase where atoms are capable of absorbing resonance
radiation before being swept out of the furnace. As the
atomization temperature is increased still further, more
of the matrix containing major amounts of nickel, cobalt
and chromium is atomized. Low energy nickel, chromium,
and cobalt absorbance lines lying within the bandpass of
the monochromator (7Å) absorb major amounts of the
continuum radiation resulting in an overcorrection error
since the indium emission from the hollow cathode lamp

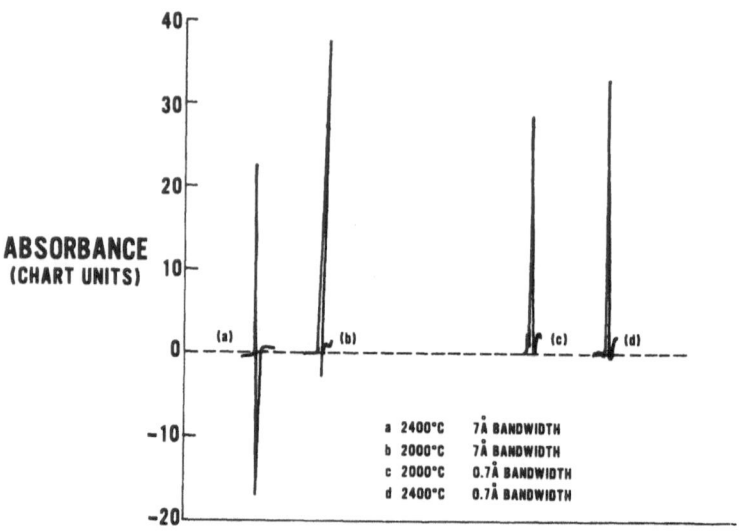

Figure 3. Effect of matrix atomic absorption on
indium absorbance.

at 3039.4Å does not overlap with the matrix absorbance
lines. The overall effect is an apparent reduction in
sensitivity for the analyte, indium, due to an overcorrection for background. This phenomenon is illustrated in
Figure 3. Recorder traces are illustrated for the atomization of indum at 2400°C and 2000°C at monochromator
bandwidths of 7Å and 0.7Å. At the 2400°C atomization
temperature and 7Å bandwidth a large negative signal results from the over-correction of the background correction
system due to atomic absorption of the continuum by
matrix atoms of nickel at 3037.9Å, cobalt at 3042.5Å and
chromium at 3039.9Å. When the atomization temperature is
reduced to 2000°C very little of the matrix is atomized,
there is little negative signal and the apparent sensitivity
is increased. When the bandwidth of the monochromator
is reduced to 0.7Å, all the matrix atomic absorption lines
are excluded, no negative signal is noted and there is
greater sensitivity recorded at an atomization temperature
of 2400°C than at 2000°C. In trace (a) the large negative
signal produces an apparent lessening of the indium ab-

sorbance signal. This type of interference can result in
significant errors when samples and standards are not well
matched with respect to absorbing elements. Atomization
temperature should be optimized to result in the maximum
analyte sensitivity at the minimum temperature possible.
This reduces background absorbance and extends the lifetime of the furnace.

SUMMARY OF CONDITIONS FOR DETERMINATION OF ELEMENTS

The following optimum conditions have been determined for
analysis for trace elements in complex alloys containing
major amounts of nickel, cobalt, chromium, aluminum,
titanium and molybdenum dissolved as described previously
in the text, except where noted. A sample volume of 50µl
is injected into the furnace. Optimum conditions were
determined with a Perkin-Elmer Model 403 atomic absorption
spechtrophotometer and HGA 2000 graphite furnace atomizer.
In many instances the identical conditions have been found
effective with the HGA 2100 atomizer.

Arsenic: Wavelength - 193.7 nm

Heating Programs   Char 1400°C - 30 sec
                    Atomize 2100°C

Lower Limit   2 ppm in alloy

Sample dissolution must be carried out in the absence of
fluoride to prevent losses. After dissolution, small
amounts of hydrofluoric acid may be added at room temperature to dissolve insoluble residue. Sensitivity for
arsenic remains approximately constant over a range of
char temperatures from 600°C to 1400°C. Sensitivity increases with increasing atomization temperature throughout
the entire range studied to 2800°C. The increased
emission intensity of the arsenic electrodeless discharge
lamp (EDL) over that of the hollow cathode lamp (HCL) is
important in achieving lowest detection limits:

Selenium: Wavelength 196.0 nm

Heating Programs   Char 1000°C - 45 sec
                    Atomize 2400°C

Lower Limit   0.1 ppm in alloy

Selenium is effectively stabilized in the furnace by con-
cominant metals throughout preatomization heating cycles.
Sensitivity decreases markedly with increasing char
temperatures above 1200°C. In materials containing large
amounts of iron a negative signal is common on atomization when a continuum source is used for background
correction. The use of the selenium EDL is recommended
for achieving lowest detection limits.

Tellurium: Wavelength - 214.3 nm

Heating Program,   Char 600°C - 60 sec
                   Atomize 2200°C

Lower Limit   0.2 ppm in alloy

At atomization temperatures in excess of 2200°C background
increases rapidly and tellurium sensitivity decreases.

Antimony: Wavelength 217.6 nm

Heating Program,   Char 400°C - 45 sec
                   Atomize 2200°C

Lower Limit   1 ppm in alloy

Background increases for this element above 2200°C, resulting in a net decrease in sensitivity. Reproducibility
is remarkably good for antimony, and a 3% coefficient of
variation has been obtained.

Bismuth: Wavelength 223.1 nm

Heating Program,   Char 800°C - 45 sec
                   Atomize 2200°C

Lower Limit   0.1 ppm in alloy

Bismuth shows only slight decreases in sensitivity on increasing char temperature from 400°C to 1000°C. The EDL
results in a more stable signal at the scale expansions
needed to reach the 0.1 ppm lower analysis limit. Monochromator bandwith should be adjusted to exclude the

222.8 nm bismuth line for maximum sensitivity

Germanium: Wavelength 265.2 nm

Heating Program, Char 1600°C - 30 sec
                 Atomize 2700°C

Lower Limit   0.5 ppm in alloy

Germanium is lost during the standard dissolution procedure. The parameters were determined by doping
solutions of alloy with a germanium stock solution.

Thallium: Wavelength, 276.9 nm

Heating Programs, Char 500°C - 60 sec
                  Atomize 2000°C

Lower Limit   0.1 ppm in alloy

For best precision, solutions for the determination of
thallium should be diluted from the standard 1g/50 ml
to 0.2g/50 ml. Thallium shows a pronounced maximum in
sensitivity with increasing char temperature (Figure 1)
at 500°C. At char temperatures less than 400°C or
greater than 600°C sensitivity rapidly decreases. The
flow of inert gas should be interrupted through the
furnace during the first 30 sec of the char cycle. Sensitivity decreases with increasing atomization temperature above 2100°C, perhaps due to the rapid increase
in background from the nickel alloy matrix.

Lead: Wavelength, 283.3 nm

Heating Programs, Char 400°C - 60 sec
                  Atomize 2000°C

Lower Limit   0.1 ppm in alloy

See Figure 1 for sensitivity vs. char temperature curves.
The presence of chloride must be avoided due to the high
volatility of lead chloride.

Tin: Wavelength, 286.3 nm

Heating Program,   Char 1000°C - 30 sec
                    Atomize 2300°C

Lower Limit   0.5 ppm in alloy

Sensitivity changes rapidly for tin and a standard
should be atomized between each sample to minimize this
effect. It is especially important to heat the tube to
its maximum temperature between each atomization in order
to avoid buildup of the analyte in the tube.

Gallium: Wavelength, 294.3 nm

Heating Program,   Char 1500°C - 30 sec
                    Atomize 2600°C

Lower Limit   1 ppm in alloy

Gallium forms a relatively stable compound in the furnace.
Relatively high char and atomization temperature should
be used for best sensitivity.

Indium: Wavelength, 303.9 nm

Heating Program,   Char 800°C - 45 sec
                    Atomize 2400°C

Lower Limit   0.5 ppm in alloy

Monochromator bandwidth should be adjusted to 0.7Å to
exclude matrix atomic absorbance lines from chromium,
cobalt and nickel. Increased sensitivity is realized
when the inert gas flow is shut off during the initial
30 sec. of the char cycle.

ATOMIZATION OF TRACE ELEMENTS DIRECTLY FROM METAL CHIPS

Recent work in our laboratory has demonstrated the feasi-
bility of atomizing many trace elements directly from
metal chips. Parameters have been optimized for the
determination of lead, bismuth, selenium, tellurium and
tin in complex nickel-base alloys. After metal chips are
machined from bulk samples, approximately 1 mg of sample
is transferred to the furnace and then atomized directly.

The absorbance signal is then compared with that of alloy standards similar in composition to the sample. There are many advantages in the direct atomization technique including the very rapid analysis time, increased sensitivity for many elements, freedom from preatomization heating losses and reduced background absorbance.

COMPARISON WITH OTHER ANALYSIS METHODS

The atomic absorption technique utilizing electrically heated furnace atomizers has proven to be the best technique currently available for the direct determination of important trace elements in complex alloys and metals which give rise to complex optical spectra. Emission spectrographic techniques using both classical excitation methods and newly developed ICP excitation are plagued with spectral interference problems in the analysis of trace elements in modern high temperature alloys. Spark source mass spectrometry has the sensitivity necessary for a direct determination of most trace elements of interest; however. the method is slow, subject to segregation effects and dependent on the availability of proper standards. Most other measurement methods require preconcentratioh of analyte or separation from interferences for an accurate measurement.

REFERENCES

1. J.Y. Marks and G.G. Welcher, Anal. Chem., 42, 1033 (1970).

2. J.A. Dean and T.C. Rains, "Flame Emission and Atomic Absorption Spectrometry," Vol. 3, Marcel Dekker, Inc., New York, N. Y. (1975).

3. J.Y. Marks, R.J. Spellman and B. Wysocki, "Effect of Non-Analyte Light Attenuation on Accuracy in Trace Element Analysis in Complex Alloys," FACSS 2nd National Meeting, Indianapolis, Ind., Oct., 1975.

4. G.G. Welcher, O.H. Kriege and J.Y. Marks, Anal. Chem., 46, 1227 (1974).

5. J.Y. Marks and G.G. Welcher, "Determination of Trace
    Elements of Metallurgical Interest in Complex
    Matrices by Non-Flame Atomic Absorption Spectroscopy,"
    ASTM E-3 Symposium on Flameless and Cold Vapor Atomic
    Absorption, Montreal, Canada, June, 1975.

TRACE METAL ANALYSIS IN CLINICAL CHEMISTRY

Douglas G. Mitchell

New York State Department of Health
Division of Laboratories & Research
New Scotland Ave., Albany, NY 12201

Trace metals are attracting considerable attention in the biological sciences. Metals such as lead, cadmium, mercury and arsenic are toxic if absorbed in moderate amounts, and they may even be deleterious to health at the background levels found in relatively polluted urban and industrial areas. Other metals such as iron, chromium and zinc are essential nutrients, and there is evidence that a significant proportion of the U.S. population is ingesting less than optimum amounts of these metals.

The biologic role of the various trace elements has been extensively reviewed elsewhere; Schroeder [1] and Reinhold,[2] among others, give excellent overviews of the subject. The major elements of interest are shown in Table I. Lead is widely recognized as causing a major public health problem in the U.S. Children aged 1-6 years living in decayed urban areas are very susceptible to poisoning from lead in chipped and peeling paint and from lesser sources such as roadside dust, putty, canned foods, pencils and newsprint. Cadmium is implicated in hypertension, an epidemic disease in industrialized countries. This metal is believed to interfere with zinc-containing enzymes as a result of excess cadmium intake, or inadequate zinc intake, or both.

Iron is the most abundant trace metal in the body, and iron deficiency anemia (often in association with lead poisoning) is an exceedingly common nutritional deficiency. Zinc deficiency impairs both wound-healing and mobilization of vitamin A from the liver. Chromium deficiency results

Table I

Trace metals levels in man, primitive and modern

Element	Primitive man (ppm)	Modern man (ppm)	Principal cause of difference
Essential			
Iron	60	60	–
Zinc	15	15	–
Copper	1.0	1.2	Copper pipes
Manganese	0.4	0.2	Refined foods
Chromium	0.6	0.09	Refined sugars and grains
Nonessential, usually toxic			
Lead	0.01	1.7	Motor vehicle exhaust, paint, solder
Aluminum	0.4	0.9	Food additives
Cadmium	0.001	0.7	Refined grains, water pipes
Mercury	<0.001	0.19	Fungicides

Adapted from Schroeder, "The Trace Elements and Man" (ref. 1).

in impairment of glucose tolerance factor, a cofactor with
insulin needed to metabolize glucose, and is associated
with atherosclerosis.

The analysis of trace metals in biological materials
is a difficult field for several reasons:

(a) Except for sodium, potassium, calcium and magnesium,
metals of clinical significance are usually present at less
than one part per million. This is near the detection limit
obtained with conventional atomic absorption spectrometry
(AA), thus limiting the use of this excellent technique.
For some applications, such as the determination of chromium
in serum, flameless AA is the most practical approach, but
these newer procedures have important limitations.

(b) Biological matrices are "difficult." Metals in
whole blood, serum and urine can often be determined without
extensive sample pretreatment; but tissue, hair and bone
samples usually require digestion or ashing. Such operations
are slow and contamination-prone.

(c) Metals such as lead, cadmium, mercury, zinc and
copper are common environmental pollutants, and samples can
be readily contaminated.

(d) Biological fluids are not stable. For example, it
is more difficult to extract metals from aged blood than
from fresh blood.

(e) The field is not well developed. Ideally an
analytical technique should be thoroughly evaluated, colla-
boratively tested and validated by long usage in many laboratories. This has not yet happened for biological trace
metals, mainly because many analyses are now being carried
out using instrumental techniques which have only become
available in the last few years.

Since the field is too large to cover adequately in a
short presentation, I propose to briefly discuss three topics:
sample contamination, solvent-extraction AA procedures and
micro AA procedures. Sample contamination is a fundamental
problem and must be considered with all procedures, particularly microprocedures. Solvent-extraction AA procedures
are well developed, and I will only take note of some
practical problems. The discussion of microprocedures

will emphasize our research work with cup/nitrous oxide-
acetylene flame instrumentation.

SAMPLE CONTAMINATION

Sample contamination is an important problem at all
stages: collection, storage, treatment and analysis. We
have had considerable experience with this problem, mainly
because of less than ideal working conditions. Some examples:

Lead is typically found in blood at 0.2 µg/ml, but
sweat can contain 1.5 µg/ml. Blood collected by fingerstick
from a dirty finger can have an apparent lead level of 1.2
µg/ml and a true level of 0.2 µg/ml. This false result is
not only clinically useless but also dangerous, since it is
customary to begin treatment for lead poisoning at 0.8 µg/
ml.

Glass collection devices contain sufficient zinc to
make them useless for blood and urine if a zinc analysis is
required.

Laboratory dust contains high levels of aluminum, iron,
copper and lead. In our laboratory we can rub a finger
across any flat surface, transfer the dust collected into a
Delves cup and produce an off-scale lead reading.

A mercury contamination problem was traced to a generally high background mercury level from sodium tetrachloro-
mercurate, which was being used as a reagent for sulfur
dioxide analysis.

Lengthy tissue digestion procedures tend to yield un-
reliable data because of high metal levels in air particulates.

In general, contamination problems are not easily quantified and can arise from unexpected sources. Risks can be
minimized by using: (a) Clean air. Air should be filtered,
preferably with electrostatic dust precipitation. (b) Clean
labware. All containers should be washed and rinsed with
deionized water or an ultrapure reagent and stored if necessary in a sealed container. (c) Clean reagents. The main
problem is obtaining ultrapure reagents. Nitric, hydrochloric, perchloric and sulfuric acids and water can be further

purified in the laboratory using a simple nonboiling distillation apparatus.[3] (d) Clean samples. Contamination of
samples in the field can be minimized, although not eliminated, by using stringent collection techniques. For
example, before fingerstick blood samples are collected,
the skin should be coated with collodion film, which forms
a barrier between the specimen and sweat, dirt and skin
debris.

Even with these precautions, sample contamination must
be considered at all times. Analytical procedures must be
validated using stock laboratory reference samples, such as
pooled blood and NBS Standard Reference Materials, carried
through the entire procedure. Sample collectors should be
trained in contamination-free procedures, and collection
materials should be monitored. Finally, it is desirable
to collect duplicate specimens. If the first specimen
shows an elevated level, the second specimen should be
analyzed.

SOLVENT-EXTRACTION ATOMIC ABSORPTION PROCEDURES

Dithizone was once widely used to chelate metals and
extract them into an organic solvent for colorimetric or AA
determination. This procedure requires some skill, and
dithizone has been almost entirely replaced by ammonium
pyrrolidinedithiocarbonate (APDC) or ammonium diethyldithio-
carbonate (DDC). These reagents can be used separately[4]
or in combination[5] to extract a wide range of metals, including first-row transition metals and other biologically
important metals such as arsenic, selenium, cadmium, tin,
mercury and lead (Table II). The extraction is pH-dependent,
but a number of metals can be determined simultaneously.[5]

The procedure is essentially: Buffer to a typical pH
of 5.0. Add chelating agent. Mix. Add organic solvent,
usually methyl isobutyl ketone (MIBK). Mix. Centrifuge.
Determine metal in the organic phase by AA.

This is a very good procedure and, in general, the
method of choice for determining trace metals in macrosam-
ples. Nevertheless, biological samples can produce more
serious problems than with simpler matrices. Here are
some of the problems which can occur:

Table II

pH range for formation and extraction of biologically important APDC-metal complexes

Element	pH for formation	pH for extraction
Chromium	2–9	3–7
Manganese	2–12	4–6
Iron	0–14	1–10
Nickel	1–14	1–10
Copper	0–14	0–14
Zinc	1–14	1–10
Cadmium	0–14	0–11
Mercury	0–14	0–11
Lead	0–14	0–8

Metals chelate over a limited pH range, and the chelating agent is unstable in strong acid solution. This method
is therefore not easy to use after acid digestion. A
difficult pH adjustment is required, with consequent sample
dilution.

APDC solutions are unstable, and some batches of reagent do not work.

Addition of phosphate (from buffer solution added at
blood banks) reduces apparent metal concentrations by about
20%. This loss does not occur in the presence of citrate
(from citric acid-sodium citrate-dextrose solution, which is
sometimes but not always added at blood banks).

Metals cannot be efficiently extracted from aged blood.
For example, blood samples stored at 20°C for 15 days show
up to 70% lead losses.[6]

Lead added to blood from some donors cannot be recovered. This happens occasionally for no apparent reason.

Inefficient shaking results in poor recoveries, with typical losses of 20-30%.

In short, solvent-extraction AA procedures can work well with biological samples, but lack of attention to detail can cause gross errors, typically 50% or more. These errors will be clinically significant in diagnosing lead and cadmium poisoning.

MICRO ATOMIC ABSORPTION SPECTROMETRY

Procedures for determining metals in microsamples by AA have been a major research focus for the last few years. There are several instrumental variations: carbon rod, graphite and tantalum strip furnaces, and Delves-cup flame instrumentation.[7] These techniques offer the advantages of minimal sample requirements and excellent detection limits, particularly absolute detection limits. On the other hand, the flameless furnace techniques are slow, require some manipulative skill, and are interference-prone.

These techniques have two important applications in clinical chemistry: (a) Analysis of fingerstick blood samples. The Delves-cup flame technique is widely and successfully used for determining lead in about 100 µl of blood. (b) Ultratrace analysis. Serum contains, for example, about 1 ng of chromium per ml, and the flameless AA procedures are much more convenient for chromium determinations (though probably less accurate) than emission spectrography and neutron activation analysis.

In addition to these applications, cup-flame procedures are valuable for improving overall efficiency, since it is possible to carry out wet digestion and ashing in the cup. This is much more efficient than using macrosample treatment procedures.

We regard the cup-flame technique as potentially the most promising micro-AA procedure because the dispensing of samples into cups permits off-line sample pretreatment. Also, the nitrous oxide-acetylene flame is a particularly

Figure 1.   Instrumentation for cup-flame micro atomic absorption spectrometry.   General view showing burner, absorption tube, cup and cup injector.

efficient atomizing medium.   Providing the sample is volatilized into the flame, metals of analytical interest should
be efficiently atomized.   This procedure should be much less
interference-prone than flameless procedures, where samples
are volatilized into an inert atmosphere.

The main drawback of the Delves-cup flame technique is
the fact that with the air-acetylene flame it can be used
to determine only volatile metals such as lead, cadmium,
arsenic and selenium.   We have overcome this problem by
using a hot nitrous oxide-acetylene flame[8] with the new
instrumentation shown in Fig.1.   The major improvements
compared with present Delves instrumentation, are: (a) A
very safe burner.   With a 2-cm^3 mixing chamber and a
capillary-tube flashback trap, explosive flashback is almost
impossible.   (b)   Use of a silicon-carbide absorption tube
instead of alumina or quartz.   (c) Use of silicon-coated
molybdenum cups.   These withstand the hot flame, acids and
oxidants.

The nitrous oxide-acetylene flame gives substantially
improved performance, compared with the cooler air-acetylene
flame, for the following reasons:

Cup temperature.   The nitrous oxide-acetylene flame
heats the cup to an equilibrium temperature of ca 1900°K,
compared with 1500°K for the air-acetylene flame (Fig. 2).
This temperature is sufficient to volatilize compounds of
all common metals except those forming refractory oxides.

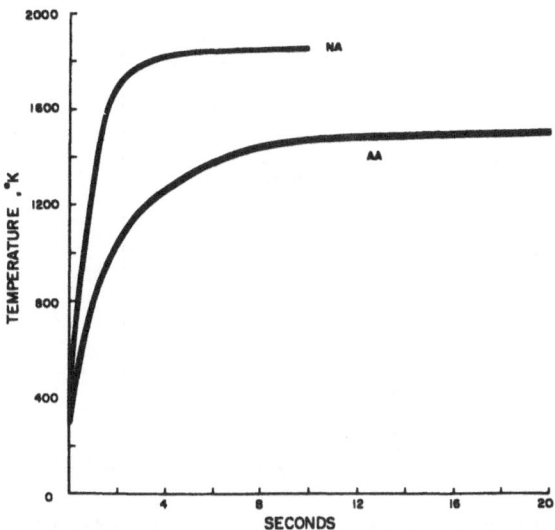

Figure 2.   Heating rates of microsampling cups calculated
from the measurement of black–body radiation from the cup
at 550 nm.   From Ward, Mitchell and Aldous (ref. 9).

Rapid sample volatilization.   The rates of volatiliza-
ion of various lead salts, calculated using latent heat and
apor pressure data, are shown in Fig. 3.   At the cup tem-
eratures reached in the air–acetylene flame, lead halides
olatilize rapidly; but PbO, hence $Pb(NO_3)_2$, volatilizes
lowly.   This at least partly accounts for observed changes
n peak shape with varying sample matrices, as well as the
nterference effects observed with peak absorbance measure-
ents.   With the nitrous oxide–acetylene flame, all lead
ompounds volatilize rapidly, and the resulting signals
hould be much less matrix–dependent.

Interferences.   It is reasonable to hypothesize that if
he cup is sufficiently hot to volatilize the sample, the
ot reducing nitrous oxide–acetylene flame will efficiently
nd reproducibly atomize most metals.   If so, the technique
hould be relatively free from interferences and certainly
ess interference-prone than flameless AA procedures.

We tested this hypothesis by determining lead, silver,
admium, zinc and copper in various "difficult" matrices:

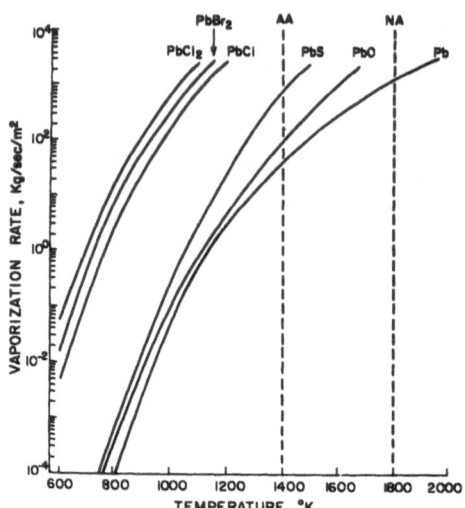

Figure 3. Computed variation of the vaporization rate of
lead compounds with increasing temperature. Maximum cup
temperatures: AA, air-acetylene flame, NA, nitrous oxide-
acetylene flame. From Ward, Mitchell and Aldous (ref. 9).

2% $AlCl_3$, 2% NaCl, whole blood, serum and lubricating oil.
Conventional AA and the cup technique with an air-acetylene
and a nitrous oxide-acetylene flame were used, and interference effects were measured from changes in calibration
curve gradients relative to the gradient of curves obtained
with aqueous standards. Gradients greater than those for
aqueous standards indicate enhancement, and lesser gradients
indicate signal suppression. Conventional AA (Table III)
is relatively interference-free, except with the lubricating
oil matrix. The cup/air-acetylene flame procedure (Table
IV) is interference-prone, but the cup/nitrous oxide-
acetylene flame procedure (Table V) shows negligible interference effects.

     Cup/nitrous oxide-acetylene flame procedures are not
yet well developed, but they show exceptional promise. The
main impediment to further development is the lack of
commercially available instrumentation. Once this becomes
available, and after further method development, I anticipate
that the cup-flame technique will be the method of choice
for applications such as the determination of copper, iron
and zinc in blood and urine and of almost all metals in
tissues.

Table III

Nebulizer/air-acetylene system: interference and precision
data

Metal	Matrix	Calibration curve gradient (arbitrary units)	RSD (%)
Lead	Aqueous	1.00	1.1
	Aluminum	0.83	6.4
	Blood	0.98	1.1
	Oil	1.53	1.1
	Salt	1.00	8.1
Silver	Aqueous	1.00	1.2
	Aluminum	0.86	3.7
	Oil	2.61	1.2
	Salt	1.01	5.3
Cadmium	Aqueous	1.00	1.5
	Aluminum	0.72	2.4
	Oil	3.22	1.3
	Salt	1.00	4.2
Zinc	Aqueous	1.00	0.7
	Aluminum	0.72	4.4
	Oil	5.36	0.8
	Salt	1.02	8.2
	Serum	1.01	1.7
Copper	Aqueous	1.00	1.2
	Oil	2.96	1.3
	Serum	1.00	2.2

From Ward, Mitchell and Aldous (ref. 9).

Table IV
Microsampling cup/air-acetylene system: interference and
precision data

Metal	Matrix	Calibration curve gradient (arbitrary units)		RSD (%) at absorbance 0.10 + 0.01	
		Absorbance		Absorbance	
		Peak	Integrated	Peak	Integrated
Lead	Aqueous	1.00	1.00	11.1	8.5
	Aluminum	0.45	0.76	13.2	8.6
	Blood	4.71	1.45	11.5	9.3
	Oil	2.02	1.41	7.7	4.6
	Salt	0.34[a]	0.38[a]	2.9	1.6
Silver	Aqueous	1.00	1.00	4.2	4.2
	Aluminum	0.54	0.85	7.3	4.2
	Oil	0.91	0.99	4.9	3.0
	Salt	0.51[a]	0.65[a]	7.3	6.2
Cadmium	Aqueous	1.00	1.00	10.9	3.4
	Aluminum	0.56	0.77	5.2	7.6
	Oil	1.33	1.01	8.0	2.9
	Salt	0.55[a]	0.66[a]	10.1	9.1
Zinc	Aqueous	1.00	1.00	8.6	8.1
	Aluminum	0.87	0.61	8.1	5.5
	Oil	1.39	1.15	7.6	2.1
	Salt	0.73[a]	0.34[a]	8.7	11.4
	Serum	1.57	1.15	9.4	4.9
Copper	Aqueous	1.00	1.00	11.1	6.1
	Oil	1.18	1.83	10.7	4.4
	Serum	6.17	3.54	9.9	5.7

[a]Significant background absorption (>0.15) was obtained
from the matrix alone.
From Ward, Mitchell and Aldous (ref. 9).

Table V
Microsampling cup/nitrous oxide-acetylene system: interference and precision data

		Calibration curve gradient (arbitrary units)		RSD (%) at absorbance 0.10 + 0.01	
		Absorbance		Absorbance	
Metal	Matrix	Peak	Integrated	Peak	Integrated
Lead	Aqueous	8.20	0.83	6.7	5.7
	Aluminum	8.09	0.83	6.6	7.9
	Blood	8.20	0.86	7.2	8.0
	Oil	8.20	0.83	8.4	8.0
	Salt	8.09	0.86	8.8	9.2
Silver	Aqueous	2.00	0.56	9.2	6.2
	Aluminum	1.98	0.54	3.4	4.2
	Oil	1.97	0.56	8.7	7.9
	Salt	2.01	0.56	6.8	6.5
Cadmium	Aqueous	2.87	0.54	4.9	5.8
	Aluminum	2.88	0.54	3.3	5.6
	Oil	2.89	0.54	6.3	3.9
	Salt	2.86	0.53	6.0	4.5
Zinc	Aqueous	3.00	0.84	6.2	5.9
	Aluminum	3.00	0.83	8.9	7.0
	Oil	2.99	0.83	8.7	8.7
	Salt	3.01	0.82	8.8	9.2
	Serum	2.97	0.82	8.6	9.9
Copper	Aqueous	20.7	5.54	8.9	8.5
	Oil	21.0	5.67	10.2	10.9
	Serum	20.8	5.63	10.5	10.2

From Ward, Mitchell and Aldous (ref. 9).

CONCLUSIONS

Trace metal analysis in clinical chemistry is difficult
both for fundamental reasons and because of special problems
related to the biological matrix. The field has undergone
spectacular development during the last five years. For
example, five years ago very few laboratories in the USA
were capable of accurate blood lead analysis, and the few
laboratories attempting serum chromium analysis obtained
data an order of magnitude too high. Now at least 50 laboratories can determine blood lead levels within about 25%,
and several laboratories can determine serum chromium.
During the next few years, increasingly reliable routine
analysis should be available, and this should enable biological scientists to better assess the very important role of
trace metals in nutrition and toxicology.

REFERENCES

1.   Henry A. Schroeder. The Trace Elements in Man. Devin
      Adair Company, Old Greenwich, Conn. 1973.
2.   John G. Reinhold. Clin. Chem. 21/4, 476 (1975).
3.   E. C. Kuehner, R. Alvarez, P. J. Paulsen and T. J.
      Murphy. Anal. Chem. 46, 1894 (1974)
4.   J. Story. "The Solvent Extraction of Metal Chelates,"
      Macmillan, New York, NY 1964.
5.   J. D. Kinrade and J. C. Van Loon. Anal. Chem. 46, 1894
      (1974)
6.   D. G. Mitchell, F. J. Ryan and K. M. Aldous. Atomic
      Absorpt. Newsl. 11, 120 (1972)
7.   H. T. Delves. Analyst (London), 95, 431 (1970)
8.   D. G. Mitchell, A. F. Ward and M. Kahl. Anal. Chim.
      Acta, 76, 456 (1975)
9.   A. F. Ward, D. G. Mitchell and K. M. Aldous. Anal.
      Chem. 47, 1656 (1975)

THE ANALYSIS OF GLASSES, REFRACTORIES, CERAMICS, AND RELATED MATERIALS BY ATOMIC SPECTROSCOPY

W. M. Wise and J. P. Williams

Research and Development Laboratories

Corning Glass Works, Corning, N.Y. 14830

INTRODUCTION

The analysis of glasses and refractory oxides has been simplified in recent years through the development of atomic absorption and emission spectrometers that have improved nebulizers, burners, and electronics. When these features are used in conjunction with the conventional excitation flames, readouts are sufficiently stable to allow the determination of some major constituents, viz., Si (1-11), alkalies (3,5-8,11-13), alkaline earths (3,5,7,8,10,11), and Al (3,5,7-11,14-16). The precision and accuracy are equal to those obtained by gravimetric and titrimetric methods. The resulting spectrometric procedures are usually faster than their more classical counterparts because arduous separations which also tend to increase blanks are no longer necessary prior to the introduction of the sample solution into the excitation source and its comparison with appropriate standards. Consequently, the major problems are now associated with adequate sample preparation and decomposition plus obviating spectral, physical, and chemical interferences. Some of these interferences can be eliminated through the use of chemical releasing agents and/or ionization suppressants (17,18), or high temperature argon plasmas (19-21); the latter being especially useful for the very refractory elements. The steepled argon plasma jet introduced by Spectrametrics, Inc. is currently being employed for determining macro and trace levels of B in glasses (22). Heated carbon excitation sources have also been found especially valuable for doing trace analyses in glasses, rocks, and minerals by atomic spectroscopy (23-26). The analyte atoms are usually vaporized from a rod or tube furnace into the optical path of the in-

strument where they absorb radiant energy being emitted
from a hollow-cathode lamp. Such methods sometimes permit
the determination of quantities of elements that are orders
of magnitude below the detection limits of flame spectroscopy or optical emission spectrographic procedures (27).

    Probably the most outstanding feature possessed by
atomic spectroscopy is the rapidity with which reliable
analytical results can be obtained. Consequently, in glasses and refractory oxides the number of major and minor
constituents being determined by this technique is continu-
ally increasing. There are sufficient comparative data in
the literature references previously mentioned to verify
that atomic spectroscopy can be used to satisfactorily determine many elements. Therefore, quantitative comparisons
in the following discussion will be chiefly limited to some
of the newer developed methods that have been found useful
for the analysis of glasses and related materials.

APPARATUS

    The instruments described below were employed to obtain
the atomic absorption and emission measurements recorded
later in this discussion. With these instruments are provided valuable manuals suggesting preliminary starting parameters for determining the various elements:

    A Varian-Techtron AA-5 spectrometer with N_2O + C_2H_2 and
    air + C_2H_2 burners, a model 70 automatic wavelength
    scanning device, plus a D-1-30 digital integrator.

    A Perkin-Elmer 403 spectrometer with N_2O + C_2H_2 and air
    + C_2H_2 burners, an HGA-2000 controller with a graphite
    tube furnace, and a deuterium arc power supply.

    Spectrametrics Spectrascan spectrometer with the argon
    plasma jet excitation source.

EXPERIMENTAL

Sample Cleaning

    In addition to being representative the sample must be
clean, particularly if trace analysis is to be done. Any

history concerning the sample may be helpful for determining the procedure to be used. Organic materials can be removed with a suitable solvent like trichloroethylene or xylene while dirt and inorganics are generally removed with water or acids such as HF and/or HCl. Then the sample usually is dried at 120°C for 1 hour.

Sample Preparation

If the sample is in large pieces, it is first crushed to a suitable size for grinding. Steel, agate, or alundum mortars can be used provided the material of construction does not contain some trace element being determined. For trace analysis involving acid decomposition of samples, the materials are crushed between sheets of clean plastic and left around 5-mesh; because grinding can often introduce serious contamination. In order to obtain rapid sample decomposition when the analyte is a major constituent, the grinding is continued to promote the entire sample through 100-mesh.

Finely divided materials that are higher in alkalies or alkaline earths may readily pick up water and carbon dioxide and should be stored in closed containers over BaO or some other suitable desiccant.

Sample Decomposition

Usually, the analyte in solution is introduced into the excitation source in the form of an aerosol. It is desirable that the solution be as free of diverse materials as possible to minimize the number of interferences and burner problems such as clogging. Since excess acids can be removed from the system by fuming, these reagents are preferred for dissolution. Sometimes a glass will dissolve with a simple acid treatment, viz., a high PbO glass with dilute nitric acid. However, in the majority of cases decomposition is not that simple.

HF + $HClO_4$. Most 0.1-g samples of glasses, glass-ceramics, and refractory oxides require a harsh treatment to effect dissolution such as heating in a platinum vessel with a mixture of HF and a high-boiling acid like $HClO_4$. The $HClO_4$ is employed for a number of reasons: once the decomposition is completed, subsequent strong fuming removes Si and B, thereby reducing the amounts of solids introduced

into the excitation source. In addition, the excess fluoride is expelled converting insoluble fluoride salts to the
very soluble perchlorates. Fluoride should also be removed
to prevent attack on glass pipets, dilution vessels, and
burners. Fuming to dryness should be avoided to prevent
the formation of solids that are subsequently difficult to
dissolve. The formation of insoluble TiO_2 can be prevented
by adding an equal molar amount of $ZrOCl_2$ before fuming.

$HF + H_2SO_4$. Because of its higher boiling point, fuming with H_2SO_4 is more effective than $HClO_4$ for removing
fluoride when the sample is high in Al. Also, when the
system contains large amounts of K, the H_2SO_4 is preferred
over $HClO_4$. However, the formation of the insoluble sulfates of Ba, Sr, and Pb can cause the loss of sample by
spattering.

HF at Elevated Temperatures. Enclosed Paar bombs containing HF at 140°C are useful for decomposing samples containing highly refractory oxides such as Al_2O_3. After dissolution, volatiles are removed by transferring to a platinum dish and fuming with $HClO_4$ or H_2SO_4.

Sampling by Chemical Etching. It is often of interest
to sample or obtain composition profiles at or near the
surface of laminated glass and glass-ceramics. If the surface glass has an expansion that is lower than the core, the
surface is under compression. Because of the resulting high
strengths, such materials are being used for an increasingly
larger number of purposes. Sampling of a surface 50 to 60 μ
thick by mechanical grinding is difficult to control. However, by the proper choice of acid mixtures, temperatures
and etching times, the depth of etch can be satisfactorily
controlled (28). The thickness of the surface layer removed can be estimated by calculation if the weight loss,
surface area etched, and density of the glass are known.

Fusions. Fusions should be avoided unless absolutely
necessary, viz., when acid dissolution may volatilize some
analyte, or if the sample is not satisfactorily decomposed
by acids. A fusion very often will introduce a large amount
of salt into the excitation source, and this may increase
the background. Also, the fluxes available to conduct
fusions are sometimes contaminated resulting in high blanks.
However, if a fusion is necessary it should be done in a
properly selected container at the lowest temperature re-

quired to decompose the sample with about a 5 to 10:1 ratio
of anhydrous flux to sample. Since most glasses and glass-
ceramics are acidic in nature, the best fluxes are basic.
Sodium carbonate in platinum at 900-1000°C is an excellent
flux for decomposing materials high in SiO_2. For more refractory samples, $LiBO_2$ or mixtures of Na_2CO_3 and $Na_2B_4O_7$
may be required. Unless absolutely necessary, fusions that
introduce significant amounts of the container, e.g., NaOH
in iron or nickel, or Na_2O_2 in zirconium, are not used for
obvious reasons. Cooled fusion melts are generally dissolved with stirring in hot 3 $\underline{M}$ acids like HCl, $HClO_4$, or
HNO_3. Sometimes the melts must be dissolved in solutions
containing complexing agents like tartaric acid to prevent
the precipitation of Nb and Ta oxides.

Preparation for Measurement

In order to obtain the best possible result, the concentration of the analyte should be adjusted by dilution to
fall within the optimum range of the calibration curve. If
bracketting standards are used, the range should be sufficiently small where the curve is, for all practical purposes,
linear.

Standards

Standard solutions are usually prepared from highly
pure oxides, carbonates, or metals using an acid dissolution
or a basic fusion. For best results, the compositions of
the standards should be adjusted to simulate the compositions of the sample solutions to compensate for any enhancement or depression of the analyte's response by diverse ions
or molecules.

Selection of Excitation Source and Mode

With flame excitation sources, the selection of the
fuel and oxidant and their ratios plus the operational mode
to be used for a particular element are dependent upon which
combination gives the most stable and sensitive signal with
the lowest background. See Tables I and II. The flame
emission mode is used for determining the alkalies, Ca, Sr,
Ba, La, Nd, and Al. On this mode the air + C_2H_2 (2300°C)
is employed for the alkalies; while N_2O + C_2H_2 (2900°C) is
used for Ca, Sr, Ba, La, Nd, and Al. The atomic absorption

Table I.   Flame Emission Excitation Conditions for Various
Elements.

Fuel Mixture	Element	Wavelength nm	Oxidant: Fuel Ratio	Calibration Range; ppm
Air	Li	670.8	Stoichiometric	0.25 – 2
+	Na	589.0	Stoichiometric	0.25 – 4
C_2H_2	K	766.5	Stoichiometric	0.25 – 4
	Rb	780.0	Slightly Reducing	0.5 – 5
	Cs	852.1	Slightly Reducing	0.5 – 30
N_2O	Ca	422.7	Oxidizing	0.25 – 3
+	Sr	460.7	Oxidizing	0.25 – 5
C_2H_2	Ba	553.5	Oxidizing	0.5 – 5
	La	441.7	Stoichiometric	2 – 25
	Nd	660.8	Oxidizing	5 – 100
	Al	396.1	Stoichiometric	0.5 – 25

Table II.   Flame Atomic Absorption Conditions for Various
Elements.

Fuel Mixture	Element	Wavelength nm	Oxidant: Fuel Ratio	Calibration Range; ppm
Air	Cr	357.9	Reducing	1 – 10
+	Mn	279.5	Oxidizing	0.5 – 5
C_2H_2	Fe	248.3	Oxidizing	1 – 10
	Co	240.7	Oxidizing	1 – 10
	Ni	239.0	Oxidizing	1 – 10
	Cu	324.7	Oxidizing	1 – 10
	Zn	213.9	Oxidizing	0.2 – 2
	Pb	217.0	Oxidizing	2 – 20
N_2O	Mg	285.2	Oxidizing	0.1 – 1.5
+	Ca	422.6	Oxidizing	0.25 – 3
C_2H_2	Sr	460.7	Oxidizing	0.5 – 5
	Al	309.2	Stoichiometric	1 – 25
	Si	251.6	Oxidizing	10 –100
	Y	410.1	Slightly Reducing	10 –100

mode is utilized for determining Cr, Mn, Fe, Co, Ni, Cu, Zn, Pb, Mg, Ca, Sr, Al, Si, and Y. On this mode air + C_2H_2 is employed for Cr, Mn, Fe, Co, Ni, Cu, Zn, and Pb; and N_2O + C_2H_2 is used for Mg, Ca, Sr, Al, Si, and Y. The following are two specific examples of how certain phenomena can affect decisions concerning procedures. The absorbance for Mg at 285.2 nm is higher by a factor of 2 or 3 with the air + C_2H_2 than with the oxidizing N_2O + C_2H_2 flame. The later excitation source is preferred when the element is present as a major constituent, because the increased stability of the signal leads to better precision of the result. In the N_2O + C_2H_2 flame, the Ba emission signal at 553.6 nm is much more sensitive than the absorbance response; and if Ca is present, a CaO band also appears at that wavelength. In this case, a satisfactory Ba emission value can still be obtained by scanning the wavelength region involved and graphically subtracting the CaO signal. See Figure 1.

The argon plasma jet provides an excitation source with a temperature of about 9000°K. At this temperature, oxides and carbides of the more refractory elements then can be determined at lower levels by emission with better accuracy and precision than by flame methods (22).

Interferences

In addition to the previously discussed spectral interferences, other kinds can present some formidable problems in flame excitation sources.

Chemical. If the energy of the excitation source is much more than required to produce the transition corresponding to the wavelength being observed, then a significant proportion of analyte atoms may become involved in ionization transitions, resulting in a decreased signal. Ways to obviate this involve making the flame cooler or adding to the solution an ionization suppressant, i.e., a substance that has an ionization potential that is relatively low and will make the flame electron rich. Potassium, Rb, Cs, and La salts will accomplish this purpose.

Some substances effect depression of a spectral signal by chemical associations which reduce the number of free analyte atoms in the flame. Aluminum and anions like phosphate and sulfate have been shown to do this with alkaline

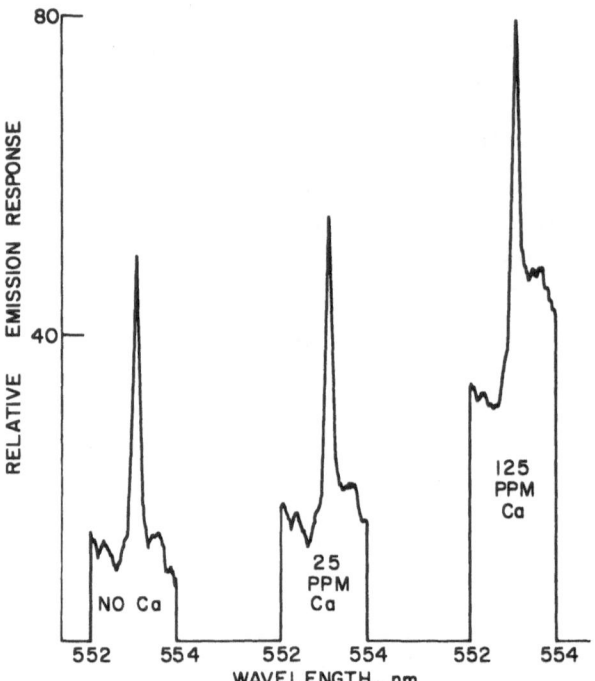

Figure 1.   Interference of the CaO Bandhead on a 2.5-ppm
Ba Response.

earths. This situation is remedied by adding large amounts
of La to tie up the anion and a complexing agent like EDTA
to cancel the effect of the Al (18).

   Physical. In addition to the chemical, certain physical changes in the sample solution can seriously affect the
result. The introduction of an organic solvent or anything
that appreciably alters the viscosity of the solution will
produce a change in the rate at which the sample goes into
the flame. Frequently, organics will also change the temperature of the flame. High salt concentrations can effect
changes in the signal by the scattering of radiant energy
and the build-up of solids on the burner slot. Sometimes
the scattering can be eliminated through the use of hydrogen
or deuterium background correctors or by wavelength scanning. Probably the best procedure for minimizing the effects of these interferences is to perform matrix matches
between samples and standards.

Data Processing

The calculation of analyte concentration is based on a
comparison of the sample solution's absorption or emission
reading with known standards. Each digital readout is the
result of an 8 to 10-sec integration time. Readings are
taken on a series of intermixed samples and standards; and
this process is repeated at least two more times, depending
on the stabilities of the signals involved, to acquire the
necessary degree of accuracy in the final result. These
data, along with the pertinent information describing sample
volume and weight, dilution and gravimetric factors, are fed,
via a remote terminal, into a General Electric computer with
a Mark II system using basic language. The 2000-data line
in Table III gives the number of standards followed by the
number of readings. Then are shown how many samples and the
gravimetric factor for $K \rightarrow K_2O$ (1.205). The rest of the line
is for the user's records. The 2001 line begins the calibration data which are four standard readings followed by
that standard's concentration and continues through 2003.
Lines 2004 and 2005 show sample identification, volume,
dilution factor, sample weight, number of readings,followed
by the individual readings for that sample. The computer
then utilizes a canned General Electric program known as
"Polfit" to show how far each calibration point is from the
curve in % difference. This is followed by a reporting of
the calculated % K_2O for each sample.

Determination of Some Important Glass Components

Silicon. Since this element is a primary constituent
of most glasses, it must be determined quite frequently.
The classical method is a gravimetric one which involves the
dehydration, separation, and weighing of SiO_2; and this is
time-consuming and costly. Now an atomic absorption method
is in use that can be applied to a wide variety of glasses,
ceramics, and related materials (29). It involves the decomposition of the sample by fusion with a mixture of Na_2CO_3
and $Na_2B_4O_7$, and dilute nitric acid is used to dissolve the
cooled melt. Silicon is then kept in solution by complexing
with molybdate at pH 1.4. A slightly oxidizing $N_2O + C_2H_2$
flame is used as the excitation source to make the measurements at a wavelength of 251.6 nm. The samples listed in
Table IV were analyzed for SiO_2 using this method. Comparison of the results with the certified and gravimetric

Table III. Computerized Calculation of % K_2O.

2000 Data 6,4,2,1.205,"WW","WWX","K","8/7/75"
2001 Data 185,193,185,187,1,237,235,236,237,1.4
2002 Data 307,303,306,308,2,401,396,397,401,3
2003 Data 478,476,478,477,4,545,540,543,544,5
2004 Data NBS-89,500,8.33,0.1014,4,271,267,272,273
2005 Data NBS-91,100,15.62,0.1011,4,279,280,279,280,11111

Calibration Data

Deg.:	Variance:	C o e f f i c i e n t s		
3 :	1.07E-04:	-.1271E+00:	.5254E+00:	.2924E-01:
				.8786E-02
2 :	2.79E-04:	.2169E+00:	.1979E+00:	.1254E+00:
1 :	3.57E-02:	-.1233E+01:	.1109E+01:	

Which Degree of Polynomial Do You Want (Type 1,2,or 3)? 3

Analysis

Observed	Calculated	Difference	% Difference
1	1.00062	6.23614E-4	6.23614E-2
1.4	1.39326	-6.74064E-3	-0.481475
2	2.00623	6.23322E-3	0.311661
3	2.99002	-9.98315E-3	-0.332772
4	4.00157	1.57195E-3	3.92988E-2
5	4.99481	-5.18894E-3	-0.103779

Sample	ppm	Volume	Dilution	Weight	% K_2O
NBS-89	1.6842	500	8.330	.1014	8.34[a]
NBS-91	1.7617	100	15.620	.1011	3.28[b]

Certificate values: [a]8.40, [b]3.25

---

values shows that the agreement is good even when elements
that form insoluble molybdates are present. In order to
avoid burner clogging the precipitates were allowed to
settle, and only the supernates were aspirated into the
flame. The NBS samples listed in Table V were also analyzed
to affirm that in addition to glasses this atomic absorption
method is useful for determining SiO_2 in other diverse materials. Probably the chief advantage is the reduction of the

Table IV. Atomic Absorption (AA) Results for SiO_2 in Glasses.

Sample Type	SiO_2	
	AA	Gravimetric
Pb-Ba, NBS-89	65.6[a]	65.4[b]
Opal, NBS-91	67.8[a]	67.5[b]
Borosilicate, NBS-93	80.3[a]	80.6[b]
Ca-P, Opal	75.9[c]	75.7[d]
Zn-F, Opal	60.4[c]	60.5[d]
Ba-Pb, TV Glass	66.2[c]	66.2[d]
Pb Glass	42.3[c]	42.0[d]
Al-Ti-Zr, Ophthalmic	58.5[a]	58.3[a]
Ba-Ti-La, Ophthalmic	21.0[a]	21.4[a]

[a]Average of two determinations; [b]NBS certified value; [c]Average of three determinations; [d]Average of five labs.

Table V. Atomic Absorption (AA) Results for SiO_2 in NBS Materials.

Sample Type	% SiO_2	
	AA*	Gravimetric
Ca-Al, Limestone NBS-1A	14.4	14.1
Al-K, Feldspar, NBS-70A	67.4	67.1
Al-Ti, Refractory, NBS-78	21.2	21.7
Al-Ti, Refractory, NBS-77	32.8	32.4
Al, Flint Clay, NBS-97A	43.6	43.7
Al, Plastic Clay, NBS-98A	48.7	48.9
Al-Na, Feldspar, NBS-99	68.5	68.7
Silica Brick, NBS-102	93.8	93.9

*Average of three determinations.

time of analysis. Six to eight samples can be analyzed in 4½ hours while the gravimetric procedure requires 1½ days.

Preliminary work has shown that if Al is present in the sample, it can also be determined in this molybdate matrix.

Alkalies. Probably the group of elements determined most frequently in glasses and refractory oxides by atomic

spectroscopy is the alkalies: Na, K, and Li. It is for-
tunate that satisfactory flame methods are available for
these elements, because the other procedures are so much
more time-consuming. The stabilities of the alkali resonance responses are about the same using either the flame
emission or the atomic absorption mode. Therefore, to eliminate another variable, viz., a hollow-cathode lamp, the
alkalies are done by emission.

     In the air + C_2H_2 flame the Na and K responses at 589.0
and 766.5 nm, respectively, are not as sensitive as in air +
natural gas due to ionization; however, in stoichiometric
air + C_2H_2 they are more stable. Since the more stable
response is the requirement for determining a major constituent, the air + C_2H_2 flame is used in conjunction with an
ionization suppressant. The use of a cooler more reducing
air + C_2H_2 flame to decrease ionization produces a luminescent flame sometimes causing instability of the signal due
to background noise. Lithium does not appear to be ionized
in an air + C_2H_2 flame, because the addition of NaCl or KCl
does not significantly change the emission response at 670.8
nm. The Na response at 589.0 nm is increased only slightly
by the addition of an ionization suppressant such as KCl,
while the K response at 766.5 nm is drastically affected by
the addition of NaCl. See Figure 2. For Na, 100 ppm of KCl
is all that is required to reach a plateau reading. On the
other hand, a very high NaCl concentration is required to
effect a maximum K response. At such high salt concentrations and even with a Code 2408 red filter in the optical
path the K response is somewhat unstable -- probably caused
by burner clogging. The best compromise here is 1000 ppm
NaCl which gives accuracies within the limitations of flame
emission measurements.

     Alkaline Earths. The alkaline earth oxides of Mg, Ca,
Sr, and Ba are sometimes included in many glass and refractory compositions in order to produce the desired physical
and/or electrical properties. It is shown in Tables I and
II that some of these elements can be determined on either
the emission or absorption modes. In the oxidizing N_2O +
C_2H_2 flame, a flame buffer consisting of a chemical releasing agent and/or an ionization suppressant is needed for
each of these elements to prevent depressive interferences
from other ions (18). Magnesium only requires 1000 ppm
$LaCl_3$. The other three are best determined in the presence
of 1000 ppm $LaCl_3$ and 0.01 $\underline{M}$ EDTA at pH 4.

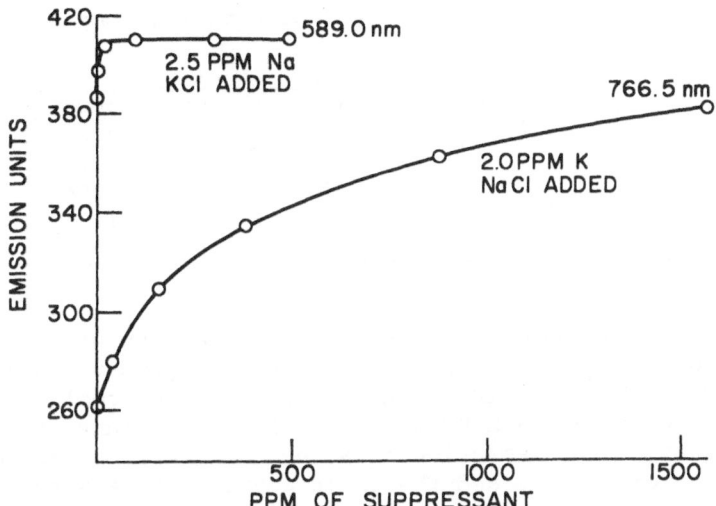

Figure 2. Effects of Ionization Suppressants on Na and K
Responses in the Air + C_2H_2 Flame.

Aluminum. The oxide of aluminum also appears in many
refractory materials and glasses. If a refractory is not
satisfactorily dissolved by acids, then a fusion with $LiBO_2$
must be used. Silicon, if present, then seriously depresses
the Al emission and absorption signals at 396.1 and 309.2 nm,
respectively, in the N_2O + C_2H_2 flame. However, this problem is eliminated by having in the standard and analytical
solutions 2000 ppm KCl and 0.01 $\underline{M}$ 1,2-diaminocyclohexane-
tetraacetic acid, all at pH 4 (16). If possible, dissolution
of the sample in HF + $HClO_4$ and fuming to remove the F^- is
preferred; but if Ca is present, the combination of Ca and
ClO_4^- produces a synergistic enhancement of the Al signal on
both operational modes. This effect can be obviated by employing a flame buffer consisting of 5000 ppm $Ca(ClO_4)_2$ and
0.12 $\underline{M}$ $HClO_4$ in both standard and analytical solutions. The
data in Table VI show the Al results for four NBS samples
done in duplicate by atomic absorption, and the agreement
with the certificate values is quite good.

Yttrium. Yttrium oxide is added to other oxides like
ZrO_2 to produce refractory materials for high temperature
applications. In order to properly stabilize the final product, close control of Y_2O_3 concentration is necessary; and
its rapid determination is considerably helpful in control-

Table VI. Atomic Absorption Al_2O_3 Results for Some NBS Samples.

| NBS Sample | % Al_2O_3 | |
	Found*	Certificate Value
70A	17.9	17.9
91	6.02	6.01
93	1.94	1.94
128	1.87	1.89

ling the manufacturing process. Such refractory materials are rather resistant to acid attack and for proper decomposition require a $LiBO_2$ + H_3BO_3 fusion. The cooled melt is dissolved in hot 3 $\underline{M}$ $HClO_4$. A sufficient amount of a mono-potassium EDTA solution is added to effect a final concentration of 0.01 $\underline{M}$, and the pH is adjusted to 4 with ammonia. Standard $\underline{Y}$ solutions are prepared by fusing known mixtures of ZrO_2 and Y_2O_3 with the same amount of the flux as was used for the samples and introducing identical quantities of the other constituents.

The Y is determined on the atomic absorption mode at 410.1 nm using a stoichiometric N_2O + C_2H_2 flame. A stastistical analysis of the results on nine samples run in duplicate and ranging from 2 to 18% Y_2O_3 shows that at the 95% confidence level the absolute error is ± 0.16% Y_2O_3. The results of some recovery studies in duplicate conducted by adding known amounts of Y_2O_3 to previously analyzed samples are shown in Table VII. These data show that within experimental error the accuracy is satisfactory.

Lead. The oxide of lead is included in glasses for certain desired optical characteristics. The determination of this element by atomic absorption at 217.0 nm with an air + C_2H_2 flame is a rather straight-forward process requiring only the use of an ionization suppressant like 200 ppm NaCl. Higher concentrations of NaCl must be avoided to prevent the precipitation of $PbCl_2$. The alkaline earths and Al do not interfere at the 100 ppm level, and a chemical releasing agent like EDTA is without effect. Usually, sample decomposition is best accomplished by treatment with HF + $HClO_4$ followed by volatilization of Si with fuming. A comparison of some lead results is shown in Table VIII. The atomic absorption values are in good agreement with those obtained by other procedures.

Table VII. Yttrium Recovery Studies by Atomic Absorption Using the $N_2O + C_2H_2$ Flame.

	Mg Y_2O_3			
Present From Sample	Added Spike	Found	Recovered Spike	% Recovery of Spike
8.40	5.00	13.4_1	5.0_1	$100._2$
8.40	5.00	13.3_6	4.9_6	$99._2$
4.82	5.00	9.8_8	5.0_6	$101._2$
4.81	5.00	9.8_8	5.0_7	$101._4$

Table VIII. A Comparison of Pb Results by Atomic Absorption (AA) and Other Methods.

	% PbO	
Type Glass	AA	Other
Pb-Ba, NBS-89	17.4_4	17.5_0[a]
Pb-Si-Ca, NBS-91	0.11	0.10[a]
Pb-Si	65.6_5	65.5_7[b]
Pb-Si	27.3_5	27.4[b]
Pb-Si-Zn	61.2	61.0[b]

[a]NBS certificate value, [b]Electrochemical result.

___

Boron. Boric oxide is included in certain heat resistant and opal glass compositions. Its determination using the $N_2O + C_2H_2$ flame and the atomic absorption mode at 249.7 nm leaves much to be desired. The stability of the B response is not as good as most elements'. Also, the sensitivity is poor requiring the use of relatively large-sized samples when the lower levels of B are being determined. These problems are the result of the high stabilities of B and oxygen or carbon combinations even at the temperature of the $N_2O + C_2H_2$ flame.

The Spectrascan Echelle Grating Spectrometer is equipped with a steepled argon plasma jet excitation source capable of producing temperatures of 8,000 - 10,000 °K. This results in stable emission readings for B at 249.8 nm plus a lowering of the sensitivity by two orders of magnitude. With this instrumentation it is possible to rapidly obtain B results that agree satisfactorily with the titrimetric mannitol method (22).

The particle size of the glass sample is reduced to
325-mesh. The sample is dissolved in 1:1 HF with cooling
to prevent the loss of boron. Suffucient KCl is added to
produce a final concentration of 2500 ppm, and the solution
is diluted to volume. The emission readings of samples and
standards are compared for calculating % B_2O_3. In Table IX
are shown some comparative data for some glass samples that
were analyzed for B_2O_3 using the atomic emission and the
titrimetric mannitol methods. The agreement is quite
satisfactory.

Trace Analysis. Heated carbon elements similar to the
graphite furnaces introduced by L'vov (30) and Massman (31)
provided the capability for determining trace quantities of
elements that are orders of magnitude below the detection
limits indigenous to the flame excitation sources. Very
often the quantity of an element required to produce a re-
spectable absorbance signal is only on the order of nanogram amounts. If Figure 3 is shown a calibration curve for
Zn. A satisfactory curve exists from a few-tenths to about
1 ng. (At higher levels of Zn the curve becomes almost
level.) In Table X are shown the lower and upper working
limits for some elements that have been determined by this
technique. Consequently, sample sizes can be very small,
viz., microliters dispensed with Eppendorf or similar pipets.

Since the standards and samples for very low level work
have such small analyte concentrations, water used for prep-
arations should be purified beyond distillation by passing it
through an ion-exchange column containing a resin like
AG®50W-X8 in its hydrogen form. Also, all laboratory ware
is thoroughly cleaned with hot 6 $\underline{M}$ HCl, then thoroughly
rinsed with deionized water before use. NALGENE® containers
are used instead of glass to avoid unwanted sorption of

Table IX. Comparison of Spectrajet and Titrimetric B_2O_3
Results.

Type Glass	% B_2O_3	
	Spectrajet	Titrimetric
Borosilicate	10.5_2	10.4_8
Zn-F, Opal	1.4_0	1.4_5
$Ca_3(PO_4)_2$, Opal	12.5_6	12.7_7

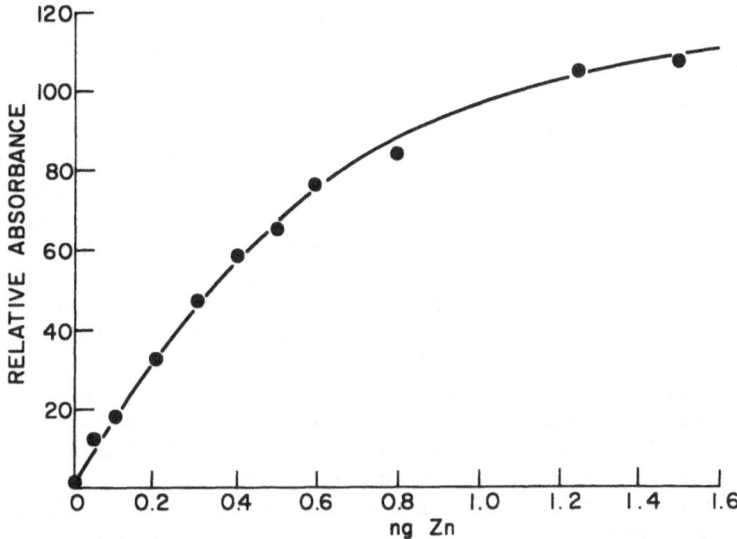

Figure 3.   Zn Calibration Curve at 213.8 nm.

Table X.   Calibration Ranges for Some Elements Determined
with a Heated Graphite Tube Excitation Source.

Element	Limits, ng		Element	Limits, ng	
	Lower	Upper		Lower	Upper
Al	0.05	5	Mn	0.01	0.5
Sb	0.4	5	Mo	0.2	15
Ba	2.5	25	Ni	0.25	10
Ca	0.05	5	K	0.05	2
Cr	0.05	5	Si	1	20
Co	0.1	10	Na	0.01	0.5
Cu	0.05	5	Sr	1	25
Fe	0.05	2.5	Sn	0.2	2
Pb	0.05	0.5	Ti	5	500
Li	0.02	1	V	2.5	50
Mg	0.01	1	Zn	0.01	1

analyte ions. Standards are prepared fresh daily and con-
tain 0.1 $\underline{M}$ HNO$_3$. Nitric acid is preferred over HCl, HClO$_4$,
H$_2$SO$_4$, or H$_3$PO$_4$ because some analyte chlorides are volatile
enough to be partially lost during charring; and some sulfates and phosphates appear to resist decomposition at the
atomization temperatures. Nitrates, on the other hand,

usually decompose through the oxide stage. The oxide is
then reduced to the element by the hot graphite, and the
analyte atoms are volatilized into the optical path at the
atomization temperature.

Sample drying, charring, and atomization times must be
empirically determined for each element placed in the furnace. The drying portion of the cycle is used to eliminate
solvents and any low-temperature boiling components. The
charring is employed to remove highly volatile materials
without loss of the analyte. The atomization temperature
is the lowest temperature which gives maximum absorbance
for the analyte. A typical experiment for estimating an
optimum atomization temperature is shown in Figure 4. The
sensitivity for Cu increases as the temperature is increased
from 2100°C to 2300°C. However, even at 2300°C all of the
Cu is not volatilized as is shown by the response recorded
when double deionized water is next injected into the furnace. At 2400°C the sensitivity for Cu is increased; and
subsequently when the water is injected, no Cu response is
observed. Atomization of samples with high salt concentrations sometimes require the use of a background corrector to compensate for the loss in radiant energy due to
scattering of the hollow-cathode's spectral line by volatilized solids. A deuterium background corrector performs this
function well between 200 and 320 nm.

In the glass industry, probably the most frequently
analyzed materials for trace amounts of harmful contaminants
are fused silica and quartz. The introduction of ppm amounts
of alkalies and alkaline earths into SiO_2 will lower its
annealing temperature and also cause devitrification. Significant amounts of first-row transition elements will produce colors, and the combination of Fe and Ti will seriously
diminish transmission in the ultraviolet region of the spectrum. Silica samples are not too difficult to analyze, because the major solid constituent is removed after dissolution and fuming with HF. For reasons of comparison several
grams of J. T. Baker ULTREX® SiO_2 was analyzed using the
graphite furnace and atomic spectroscopy and then by spark
source mass spectrometry for the elements shown in Table XI.
The certificate value obtained by optical emission for each
element is also listed. In general, the agreement is reasonably good for the levels involved.

Glasses that contain significant amounts of elements
that are not volatilized by fuming with HF can also be

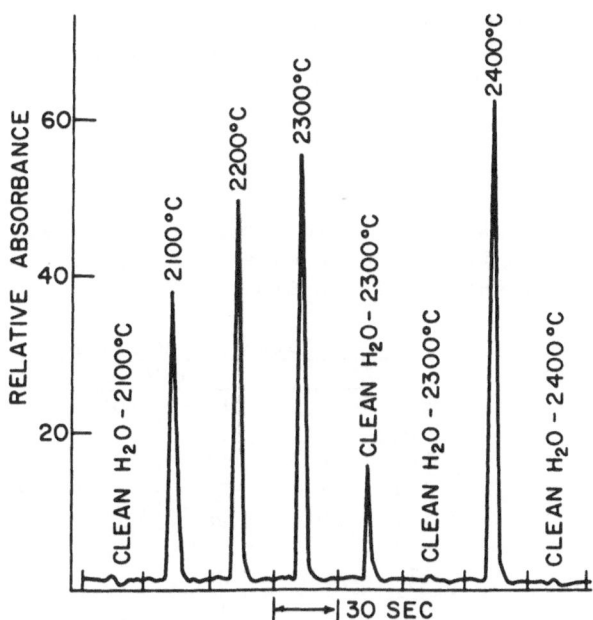

Figure 4. Effect of Atomization Temperature and the Response for 2 ng of Cu at 324.7 nm.

analyzed by atomic spectroscopy using the graphite furnace and the deuterium background corrector. Cobalt, Ni, and Cr were determined in some samples of a soda-lime ophthalmic glass; and it was established that contamination by Co at the 3 to 5 ppm level is indeed responsible for the undesirable dirty gray appearance of the glass. See Table XII.

CONCLUSIONS

Atomic spectroscopy is being applied for the determination of an increasingly larger number of elements at both the macro and sub-micro levels. Major constituents can be determined by emission or absorption with accuracies and precisions that are competitive with the older classical procedures. In addition, the analysis times necessary by atomic spectroscopy are sometimes one-third those required for the classical methods resulting in a large decrease in the cost of an analysis. Capabilities in the area of trace analysis have also been expanded by the introduction of atomic spec-

troscopy with flameless excitation sources. In some cases,
the resulting calibration curves extend from a few tenths
to several nanograms of an element.

Table XI. Comparison of Atomic Absorption (AA) with Optical
Emission (OE) and Spark Source Mass Spectrometry (SSMS)
Results on J. T. Baker ULTREX® SiO_2.

Element	Parts per Billion		
	AA	OE*	SSMS
Na	9×10^3	8.0×10^3	−
Mg	40	20	1.2×10^2
Cr	15	20	52
Mn	4.1	3	4.1
Fe	1.8×10^2	20	64
Ti	< 100	< 1	1.4×10^2
Zn	41	20	7.9
Pb	< 1	1	< 7
Li	6.9	< 100	−
K	19	< 50	−
Ca	3.9×10^2	50	4.3×10^2
Co	< 10	1	3.4
Al	20	20	4.7
Cu	4	2	4.7×10^2
Ni	6	2	29

*J. T. Baker certificate value.

Table XII. Cobalt, Ni, and Cr Results for Soda-Lime
Ophthalmic Glass.

Appearance of Glass	% T, 640 nm*	Parts per Million		
		Co	Ni	Cr
Clear Pink	74	1.0	1.0	0.4
Slight Gray	69	3.4	1.0	0.4
Very Dirty Gray	64	5.0	1.0	0.3

*In glass, Co is known to absorb at this wavelength.

REFERENCES

(1)   R. J. Guest and D. R. MacPherson, Anal. Chim. Acta, 71, 233 (1974).

(2)   J. C. VanLoon and C. Parissis, Analytical Letters, 1 (8), 519 (1968).

(3)   J. C. VanLoon and C. Parissis, Analyst, 94, 1057 (1969).

(4)   W. J. Price and J. T. H. Roose, Analyst, 93, 709 (1968).

(5)   J. W. Yule and G. A. Swanson, Atomic Absorption Newsletter, 8, 30 (1969).

(6)   F. J. Langmyhr and P. E. Paus, Anal. Chim. Acta, 43, 397 (1967).

(7)   F. J. Langmyhr and P. E. Paus, Atomic Absorption Newsletter, 7, 103 (1968).

(8)   G. H. Omang, Anal. Chim. Acta, 46, 225 (1969).

(9)   L. Capacho-Delgado and D. C. Manning, Analyst, 92, 553 (1967).

(10) Bedrich Bernas, Anal. Chem., 40, 1682 (1968).

(11) P. L. Boar and L. K. Ingram, Analyst, 95, 124 (1970).

(12) M. A. Hildon and W. J. F. Allen, Analyst, 96, 480 (1971).

(13) R. P. Eardley and R. A. Reed, Analyst, 96, 699 (1971).

(14) W. D. Cobb and T. S. Harrison, Analyst, 96, 764 (1971).

(15) P. E. Paus, Anal. Chim. Acta, 54, 164 (1971).

(16) E. C. Goodrich and D. O. Robinson, Corning Glass Works, Unpublished Work, 1974.

(17) W. Schuhkencht and H. Shinkel, Z. Anal. Chem., 194, 161 (1963).

(18) P. B. Adams and W. O. Passmore, Anal. Chem., 38, 630
      (1966).

(19) V. A. Fassel and R. N. Kniseley, Anal. Chem., 46,
      1110A (1974).

(20) E. C. Butler, R. N. Kniseley, and V. A. Fassel,
      Anal. Chem., 47, 825 (1975).

(21) G. F. Larson, V. A. Fassel, and R. H. Scott, Anal.
      Chem., 47, 238 (1975).

(22) R. A. Burdo and W. M. Wise, Corning Glass Works,
      Unpublished Work, 1975.

(23) C. W. Fuller, Anal. Chim. Acta, 62, 261 (1972).

(24) Yu. I. Belyaev, A. M. Pchelinstev, and N. F. Zvereva,
      J. Anal. Chem. USSR, 26, 422 (1971).

(25) Yu. I. Belyaev, A. M. Pchelinstev, and N. F. Zvereva,
      J. Anal. Chem. USSR, 26, 1153 (1971).

(26) H. Heinricks and J. Lange, Freseneus', Z. Anal. Chem.,
      265, 256 (1973).

(27) W. M. Wise and G. A. Machajewski, Corning Glass Works,
      Unpublished Work, 1975.

(28) D. O. Robinson, Corning Glass Works, Unpublished
      Work, 1972.

(29) R. A. Burdo and W. M. Wise, Anal. Chem. in press,
      1975.

(30) B. V. L'vov, J. Engng. Phys. 2, 44 (1959).

(31) H. Massman, Spectrochim. Acta, 23B, 215 (1968).

AN OVERVIEW OF THE DETERMINATION OF ENVIRONMENTAL CONSTITUENTS FROM THE VIEWPOINT OF A CHEMIST

Richard J. Thompson

U.S. Environmental Protection Agency

Research Triangle Park, N. C. 27711

Environmental constituents, components of air, earth, and water, are determined for various reasons. The determination of major constituents in the main presents few problems and is therefore an area for us to set aside in this presentation. We are concerned here with the determination of trace constituents, primarily in air and water, many of which are chemically and/or biologically very active. The reasons for the determinations of trace materials are many, but primarily are:

(1) To determine the levels of materials of possible adverse effect to human health. (SO_2, asbestos)

(2) To determine the nature of atmospheric chemistry by studying interactions of materials to form and to remove atmospheric components. (Ozone, NO_x)

(3) To understand changes in climate and weather caused by chemical and physical interactions. (Aerosols)

(4) To understand ecological changes induced from changing levels of substances in air, earth, and water. (SO_2 at Ducktown and Donora, volatile fluoro-

carbons)

(5)   To study changing distribution and eco-
      logical patterns in space and time.
      (DDT, PCB's)

These objectives can only be attained by definitive determinations on a continuing basis at
either background, source, or urban sites -- i.e.,
monitoring.   It is my opinion that these objectives cannot, unfortunately, be met by short-term
intensive studies, or by monitoring at single
sites, which are currently the primary modes of
investigation.   Definitive monitoring requires
continuing accurate determinations at well chosen
sites, especially in the important case of assessment of background levels.

"The more complex the sample, the more difficult the sampling; the more complex the sample,
the more difficult the analysis" are verities.
The ultimate challenge on earth to the analytical
chemist is the characterization of the constituents of air, earth, and water, to the level which
permits understanding of chemical (and biological)
processes and from them physical processes which
affect the environment.

The measurement of trace and reactive constituents in rapidly changing media, such as some
atmospheres and some waters, could in my opinion
best be accomplished by direct measurement.   Only
now is this becoming feasible for some few substances (primarily gaseous) and then, usually,
only in certain cases when concentrations far exceed background levels.

Thus in order to measure trace components of
interest (and we are almost always concerned with
traces except for evaluation of events impinging
on human health), we must collect samples and con-
centrate them either during or following the collection process.   An unfortunate consequence is
that the collection process almost of necessity
involves a change in the nature of the substance

sought unless it is a robust molecule.   For example, in collecting and analyzing solid matter
in Lake Superior in connection with asbestos, it
proved to be impracticable to sample directly on
shipboard, and storage of the samples induced
changes.   (Loss of components to container walls
and precipitation of solids is to be expected
with aqueous samples, including rain.)   In the
sampling of air, gaseous components react with
each other, with collected particulate matter,
and with the collector or filter.

We have then a picture wherein very specialized techniques of collection are needed and are
then followed by analytical determinations deman-
ding in sensitivity and in attention to detail.
The best analytical talents and instrumentation
available are needed.

However, this is not always the case.   The
amount of effort expended in the area of environmental assessmen⊥ has greatly increased since
1968 (when I left teaching), and the quality of
effort has increased somewhat due in no small
part to improvements in instrumentation.

In the laboratory, we are usually in a relatively good position; in the area of sampling
and storage, improvements must be made if long-
term trend assessments of value are to be made.
Let us take an example of the assessment of wet
precipitation (rain).   This is of interest to
soil scientists, ecologists, meteorologists, and
atmospheric chemists.   If the estimates of
Swedish workers are correct, it is also of interest to foresters and fishermen, as it is stated
that acid rain has caused the loss of trees and
of fish there.   It is of interest that the pH
of rain in the northeastern U. S. is lower than
that of Swedish rain, and therefore more acidic.

In the U. S., there are a number of workers
measuring precipitation samples, but I understand
only one (NOAA-WMO-EPA) monitoring system operates nationally.   Inasmuch as station sites in-

clude Samoa, Mauna Loa, and Fairbanks, it is
possible that 60 days may elapse between rain
collection and analysis.   Obviously this may lead
to the generation of nonsense numbers.   (One person handles the network;   it is heartening to
note that this person works full time.)

        The circumstances are clear and not atypical
-- limited manpower, primitive sampling and
storage, good chemical procedures in the lab.
I have tried to paint a picture for you of an
area which needs help from analytical chemists,
for development of better methodologies for environmental assessment, primarily in the area of
sampling.

        History gives reasons why improvements so
direly needed may be realized from analytical
chemists.   Historically, concern with terrest-
rial waters was in the purview of people oriented
toward sanitation, concern with atmospheric components was primarily centered in the ranks of
industrial health workers, consideration of sea-
water contents was the business of oceanographers,
etc., and until recently, much of the work in
many fields was done by people with limited chemical backgrounds.   The increasing overall level of
competency in analytical chemistry has been noti-
cible in the area of environmental assessment.
Many of the improvements noted can be attributed
to analytical chemists;   more effort is needed.
Your consideration of the problems faced, and par-
ticipation in seeking solutions based on sound
analytical methodologies is sought and will be
appreciated.

CHARACTERIZATION OF MUNICIPAL INCINERATOR EFFLUENTS

Stephen L. Law*

Bureau of Mines, U.S. Department of the Interior
College Park Metallurgy Research Center
College Park, Md. 20740
    and
Department of Chemistry
University of Maryland
College Park, Md. 20742

Robert R. Greenberg*

Department of Chemistry
University of Maryland
College Park, Md. 20742

The output of a municipal incinerator consists basically of three components: atmospheric emissions, aqueous effluents, and solid residues. Of primary environmental concern are the atmospheric and aqueous emissions. A thorough study of the Alexandria, Va., incinerator effluents has been undertaken with many basic analytical skills being required in the process. The following discussion will be concerned principally with sampling, sample preparation, analytical methods, and data treatment, leaving the discussion of actual results for subsequent publications.

ALEXANDRIA, VA., INCINERATOR

The Alexandria Municipal Incinerator, located at 5301 Eisenhower Avenue, serves the approximately 114,000 inhabitants of the 41-km^2 city of Alexandria, Va. It

---

*This research forms a part of dissertations to be submitted to the Graduate School, University of Maryland, in partial fulfillment of the requirements for the Ph.D. in chemistry.

operates continuously from the time it is fired up on Monday
until it is shut down early Saturday morning. Maintenance
and repairs are done Monday mornings before beginning
operation. The incinerator is designed to incinerate 270
metric tons (tonnes) of solid waste daily in two 135-tonne/
day rocking grate continuous-flow stoker furnaces. The
three rocking grates of the primary combustion chamber in
each furnace are fed through a large hopper (figure 1).
After the combustion process the hot residues are dropped
into a water quench tank. Following quenching, they are
carried by drag conveyor to a chute for loading onto
disposal trucks. The quench water is screened before going
into a holding tank from which it is recycled through the
system. Samples of this quench water were taken prior to
entering the holding tank.

Upon leaving the primary combustion chamber, the flue gases
enter a secondary combustion chamber at a temperature of
about 590° C. Before passing to the atmosphere through the
61-meter stack, the gases pass through a baffle scrubber

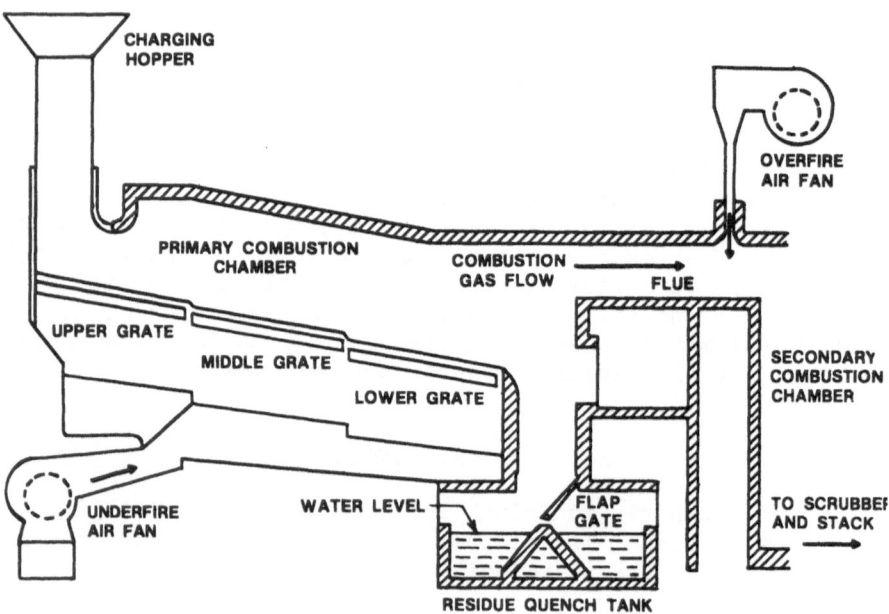

Figure 1. - Alexandria Municipal Incinerator

system where water flowing down the walls of the baffle removes fly ash that impinges on the walls. This water is then pumped to a holding tank separate from the tank used for the quench waters and recycled through a system completely separate from that of the quench waters. Heavy solids in the fly-ash scrubbing waters are allowed to settle in the holding tank and are removed by drag conveyor to trucks for disposal. Samples of the scrubber waters used in this study were taken from this settling tank.

The fly-ash scrubber water becomes highly acidic and must be neutralized during recycling to a pH between 6 and 7 by the continuous automatic addition of a "58% light soda ash." Approximately 135 kg/day of soda ash is added to the approximately 114,000-liter-capacity fly-ash scrubber water system.

The acidity of the fly-ash scrubber water is due to the various acidic gases formed during the incineration process (1). Hydrogen chloride (HCl) is formed during the incineration of chlorinated plastic materials such as polyvinyl chloride. However, it has been observed (2) that about 76 percent of the HCl emission from an incinerator comes from sources other than polyvinyl chlorides, with sodium chloride in the refuse apparently producing "very significant" quantities of HCl gas when incinerated. The incineration of sulfur-containing materials, such as synthetic and natural rubber products, contributes to the emission of sulfur oxides and sulfuric acid. Some hydrogen fluoride (HF) is also produced during the combustion of Teflon-like products and some insecticides. A study of four municipal incinerators in the New York City metropolitan area (1) revealed that, even with scrubbing facilities, the gaseous emissions contained from 38 to 113 ppm (by volume) HCl, 25 to 76 ppm H_2SO_4, 13 to 39 ppm SO_2, and 0 to 3 ppm HF. Because these acids dissolve in the fly-ash scrubber waters, these waters more readily leach metals from the ash materials and were found to be higher in dissolved metals than the quench waters.

Approximately 132,000 liters of water are used to fill the quench and fly-ash scrubber systems of the Alexandria incinerator, with only about 19,000 liters of the total being used as quench waters. The scrubber waters are recycled at a rate of approximately 2,000 liters per minute, and the quench waters at about 1,000 liters per minute.

Water losses due to evaporation and drag-out are automati-
cally replaced with charging chute cooling water to
maintain a constant water level in the separate holding
tanks. Samples of this input city tap water were also
taken. Both quench and fly-ash scrubber waters are
discharged into the local sewer system early Saturday
morning after being recycled all week, and fresh tap water
refills the systems at startup the following Monday.

The ~290° C stack gases were sampled by inserting a
probe half way up the 61-meter stack as described later.
Only particulate emissions were sampled. Previous studies
have been made of gaseous emissions from other incinerators (1, 2).

SAMPLING OF AQUEOUS SYSTEMS

Sampling sites for aqueous effluents were selected to
provide the most representative determination of dissolved
metals content possible for the aqueous effluent being
studied. For this reason stagnant areas, or areas of slow-
moving water, were avoided, and samples were taken in the
most accessible turbulent regions.

Discrete grab samples were taken throughout the study.
Flow proportional sampling was not considered necessary
because of the essentially constant flow during operation.
The samples were collected at selected intervals for
several studies; other samples were taken randomly during
daytime operation. The temperature of the waters for each
sample was taken in situ at the time of sampling. The
volume of each sample was approximately 2.5 liters except
for one composite sample of 20 liters used as a reference
solution for the various analytical methods.

Five replicate samples were taken in separate bottles
at the same location in the same time period to determine
how representative each sample and analysis would be.
Table 1 shows the reproducibility obtained on the five
replicate samples. Mercury at the part-per-billion level
is possibly adsorbing onto the fly ash particles and being
removed during filtration or adsorbing onto the container
walls before addition of the acid preservative, thus
explaining the approximately 70 percent relative standard

TABLE 1. – <u>Sampling and analytical reproducibility</u>
<u>for incinerator aqueous effluents</u>

Replicate number	Cd, ppm	Cu, ppm	Hg, ppb	Mn, ppm	Pb, ppm	Zn, ppm
1	2.7	0.49	0.8	6.0	18	130
2	2.7	.43	.5	5.7	17	120
3	2.6	.44	2.0	5.8	16	125
4	2.7	.49	3.5	5.9	17	125
5	2.6	.47	1.3	5.6	17	126

deviation for this element. The other elements in the same
samples are 100 to 100,000 times more concentrated and are
less prone to adsorption problems than mercury.

AQUEOUS SAMPLE PREPARATION AND PRESERVATION

The aqueous samples were returned to the laboratory and
filtered as soon after sampling as possible. The samples
were filtered hot, with a temperature drop of usually less
than 20° C between the time of collection and filtration.
The solutions were vacuum-filtered through 12.5-cm-diameter
Whatman No. 541* filter paper using a Buchner funnel. The
fly-ash scrubber water samples filtered fairly quickly,
requiring only one filter paper. However, some of the
alkaline quench water samples contained slimy solids which
quickly clogged the filter paper and required as much as
four changes of filter paper to filter the entire 2.5
liters. The solids from the fly-ash scrubber waters were
essentially ash material. The solids from the quench
waters contained some unburned materials such as bits of
printed paper and leaves, and even an unsinged whole maggot
was filtered from one sample.

The solids on the filter paper were air-dried for a week
or longer before being weighed. After weighing, each sample
was ground with mortar and pestle and mixed thoroughly
before samples were taken for analysis.

---

*Reference to specific products does not imply endorsement
by the Bureau of Mines.

After filtration, the hot solutions were allowed to reach ambient room temperature before a pH reading was taken. The pH of each sample was measured using a pH meter with a glass and reference combined electrode. The pH meter was calibrated before each set of measurements using pH buffers of 4.0 and 10.0.

The standard method used to preserve dissolved metals in solution is to add 5 ml of concentrated nitric acid per liter of sample, or sufficient nitric acid to lower the pH to less than 2 (3). This should prevent significant loss of dissolved metals through hydrolysis or adsorption for 6 months or longer (4). Approximately 250 ml of each filtered aqueous sample was placed in an aqua-regia-cleaned bottle, and 5 ml of 6 N HNO$_3$ were added. This quantity of acid was sufficient to lower the pH of fly-ash scrubber waters, quench waters, and other water samples to 1.5 or less. To determine the effectiveness of this preservation technique, an Alexandria Incinerator fly-ash scrubber water sample analyzed initially in July 1974 was reanalyzed 10 months later in May 1975. As shown in table II, for the metals determined no significant change occurred in concentration that could not be attributed to slight differences in standards, dilutions, or instrumental parameters. The acidified samples were used without further treatment for atomic absorption analysis, except when dilutions and/or interference inhibitors were required for certain elements.

TABLE II. - <u>HNO$_3$ (1%v/v) as a preservative for</u>
<u>dissolved metals</u>

Element	Concentration, ppm	
	July 1974	May 1975
Al	17.7	17.9
Cd	.78	.82
Cu	.20	.20
Mg	62.	60.
Mn	2.1	2.3
Pb	5.5	5.0
Zn	40.	43.

ISOKINETIC STACK SAMPLING

Samples of the atmospheric particulate emissions were collected isokinetically through a 10-cm port located midway up the 61-meter stack. The types of samples collected were both whole-filter and cascade impactor (which separates the particles by size (4)). A modified EPA-type sampling train (5), shown in figure 2, was used. The modification places the filter or cascade-impactor sampling device inside the stack instead of outside as specified by EPA. This eliminates the condensing of volatiles from the gas stream and loss of particles on the probe walls (especially important for the cascade impactor).

Also, because we were interested only in the analysis of the particulates and not in an accurate measurement of gross particle emissions, we deviated from EPA specifications by sampling at only the center of the stack. Shorter sampling times (5 to 60 minutes) were also used to prevent overloading of the cascade impactor.

An in-stack cascade impactor (6) or a stainless steel in-stack filter holder was used to collect the stack samples. An S-type Pitot tube and a pyrometer were inserted into the stack along with one of the sampling devices. While the sampling device was warming to stack temperature, the temperature and velocity of the stack gases were measured with the pyrometer and the Pitot tube connected to a 2.5-cm draft gauge. When conditions became stable a sample was taken.

Following the sampling device, the sampling train contains impingers, which condense some of the volatiles and cool the gases. Impingers 1 and 2 (see figure 2) are filled with water, No. 3 is empty, and No. 4 contains $CaSO_4$ desiccant. After the impingers is a vacuum gauge (G), followed by a coarse valve (V-1) and a fine valve (V-2) connected as a bypass across the pump. These valves maintain a flow velocity equal to that of the gas in the stack. Following the pump is a dry-gas meter used to measure the total gas volume sampled. The inlet and outlet temperatures are measured at T_1 and T_2, respectively. The outflow from the gas meter is passed through a calibrated orifice connected to a 13-cm draft gauge. Details of the use of nomograms, moisture, temperatures, pressures,

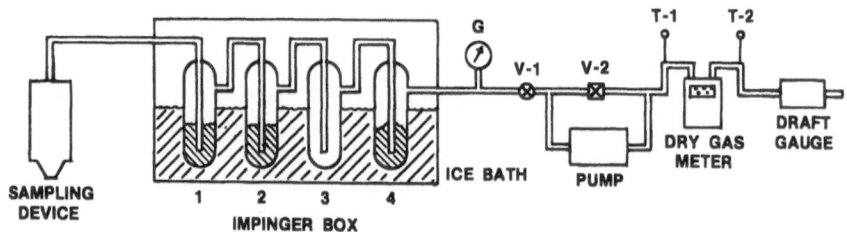

Figure 2. - In-stack sampling train

velocities, calculations, etc., for isokinetic field
sampling can be found in other references (7, 8). At the
end of the sampling period the Pitot tube and sampling
assembly are extracted from the stack and the sampling
device is cooled prior to transportation to a clean
location for reloading.

The high stack-gas temperature (∿290° C) required the
use of glass-fiber filters for whole-filter samples and
impactor back-up filters despite high blank values for
many elements. New Teflon-fiber filters with much lower
blanks are available and are presently being tried.
Impactor collection surfaces were Teflon films coated with
silicone resin to prevent particle bounce-off and re-
entrainment.

NEUTRON ACTIVATION ANALYSIS

Instrumental Neutron Activation Analysis (INAA) is an
analytical technique that can be quite effectively applied
to many types of environmental samples, possessing the
advantages of being nondestructive and extremely sensitive
for many elements. To analyze a sample by INAA, a sample
and a multi-elemental standard are irradiated together by
neutrons, resulting in radioactive isotopes of many of the
elements present. As these isotopes decay, many of them
emit γ-rays of energies characteristic of the decaying
nuclides. Energy spectra of the sample and standard are
taken, and the amount of an element in the sample can be
determined by comparing the count rates of a specific
gamma ray characteristic of the element.

INAA was used to analyze the suspended particles from the stack and fly ash from the scrubbers of the incinerator. These samples, the total mass of which ranged from less than 1 milligram to 200 milligrams, were analyzed for 35 elements. They were irradiated in the National Bureau of Standards reactor at Gaithersburg, Md., at neutron fluxes of between 1×10^{13} and 6×10^{13} neutrons-cm^{-2}sec^{-1}. Activation was done in two steps. The samples were first irradiated for a short period of time, averaging about 1 minute, and counted immediately in order to determine elements having short-lived irradiation products. A second count was then made to determine the elements with slightly longer half-lives. The same samples were then re-irradiated for a longer period of time averaging about 2 hours, and were counted after a period of several days during which the shorter-lived isotopes decayed away. Following this, another decay period and count was done to determine elements having very long half-lives. The γ-energies used for the short- and long-lived irradiation products are presented in table lll.

Standards are prepared by pipetting known volumes of standard solutions of the elements onto filter paper and allowing the paper to dry. Results using these standards were compared with results by atomic absorption for particles on the same glass-fiber filters, and good agreement was found.

Counting of the radioactivity was done using various Ge(Li) detectors ranging in size from 40 to 65 cc. With these detectors, resolutions of 1.8- to 2.2-keV full width half maximum are realized at 1332 keV, thus permitting the analysis of the complex spectra resulting from the neutron irradiation.

There is a problem involved in analyzing samples on glass-fiber filters by INAA. Glass-fiber filters are typically composed of a borosilicate-type glass. Boron has a very high cross section for neutron absorption (an exergonic process), and the sample can heat up and melt out of the polyethylene irradiation container. To get around this, samples on glass-fiber filters were irradiated for a maximum of 30 minutes. Only one 47-mm filter could be irradiated at a time.

TABLE III (A) - Short-lived irradiation products

Element	Product	γ-energy (keV)	Half-life (sec)
Ba	^{139}Ba	166	4,970
Ti	^{51}Ti	320	347
Sr	^{87m}Sr	388	10,020
In	^{116m}In	417	3,222
Zn	^{69m}Zn	439	50,040
I	^{128}I	444	1,500
Br	^{80}Br	616	1,056
Br	^{80}Br	666	1,056
Mg	^{27}Mg	1,014	568
Cu	^{66}Cu	1,039	306
Na	^{24}Na	1,368	54,000
V	^{52}V	1,434	227
Cl	^{38}Cl	1,642	2,238
Al	^{28}Al	1,779	138
Mn	^{56}Mn	1,811	9,300
Mn	^{56}Mn	2,112	9,300
Cl	^{38}Cl	2,167	2,238
Na	^{24}Na	2,754	54,000
Ca	^{49}Ca	3,084	528

TABLE III (B) - Long-lived irradiation products

Element	Product	γ-energy (keV)	Half-life (hr)
Sm	^{153}Sm	103	46.8
Hf	^{181}Hf	133	1,020
Ce	^{141}Ce	145	780.7
Yb	^{169}Yb	177	760.2
Lu	^{177}Lu	208	161.8
Ba	^{131}Ba	216	288

TABLE III (B) – <u>Long-lived irradiation products</u> – Continued

Element	Product	γ-energy (keV)	Half-life (hr)
Se	^{75}Se	264	2,890
Se	^{75}Se	279	2,890
Th	^{233}Pa	312	648
Cr	^{51}Cr	320	667.2
Cd	^{115}Cd ^{115m}In	336	53.5
Au	^{198}Au	411	64.7
W	^{187}W	479	23.9
Hf	^{181}Hf	482	1,020
La	^{140}La	487	40.2
Ba	^{131}Ba	496	288
Cd	^{115}Cd	527	53.5
As	^{76}As	559	26.4
Sb	^{122}Sb	564	64.3
As	^{76}As	657	26.4
Ag	^{110m}Ag	657	6,120
W	^{187}W	686	23.9
Cs	^{134}Cs	795	17,950
Ag	^{110m}Ag	885	6,120
Sc	^{46}Sc	889	2,013
Eu	^{152}Eu	964	102,100
Fe	^{59}Fe	1,099	1,104
Zn	^{65}Zn	1,115	5,832
Sc	^{46}Sc	1,120	2,013
Co	^{60}Co	1,173	46,070
Ta	^{182}Ta	1,189	2,760
Ta	^{182}Ta	1,221	2,760
Fe	^{59}Fe	1,291	1,104
Co	^{60}Co	1,332	46,070

TABLE III (B) - <u>Long-lived irradiation products</u> - Continued

Element	Product	γ-energy (keV)	Half-life (hr)
Eu	^{152}Eu	1,408	102,100
La	^{140}La	1,596	40.2
Sb	^{124}Sb	1,691	1,442

As one of the checks on the reliability and accuracy of atomic absorption analyses, several aqueous effluent samples were also analyzed by neutron activation techniques. Samples of 50 to 80 grams of the wastewaters were weighed in linear polyethylene containers, frozen with liquid nitrogen, and vacuum-freeze-dried in the freeze-drying facilities of the radiochemistry clean room of the National Bureau of Standards. The polyethylene containers, the freeze-drying apparatus, and the freeze-drying procedure are described in detail elsewhere (8). Less than 5 percent loss of all but the most volatile elements (for example, mercury and iodine) is reported for this preconcentration technique (8). Many of the elements producing short-lived isotopes (Table III) could not be determined in the freeze-dried samples because of the high sodium concentrations. After the interfering sodium activity had died away, however, the long-lived isotopes could be determined. A representative comparison of atomic absorption results with results from freeze drying followed by neutron activation analysis is shown in table IV. Except for K and Zn, the concentrations shown were close to the detection limits for atomic absorption.

TABLE IV. - <u>Comparison of freeze-drying</u>
<u>INAA with AA analysis</u>
<u>of aqueous effluents</u>

Element	AAS	INAA
Ag	0.2	0.13
Al	5	4.8
Fe	.7	.78
K	44	49
Zn	103	106

FLAME ATOMIC ABSORPTION

The popularity of atomic absorption spectrophotometry can be attributed to its moderate cost, useful levels of detection limits, precision, and accuracy for about 65 elements, the ease of aqueous sample preparation, and the simplicity of operation with essentially no "interpretation" of result for the user to learn. However, its apparent simplicity can be deceptive. At times faulty data may be reported because improper flame conditions were used, ionization, broad-band absorption, or flame-scattering problems were not considered, or depression or enhancement by other ions was not anticipated. Table V shows the possible interferences and the steps taken to eliminate these interferences for elements determined in the samples of this study. Also shown are the type of flame used and the sensitivities for the different elements under ideal conditions. Lanthanum was not used for ionization suppression as recommended in many methods manuals because a sufficiently pure lanthanum compound was not available. Significant quantities of such elements as Ca, Cu, Fe, Mg, and Zn were found in the bottles of 99.997 percent La_2O_3 initially purchased for ionization suppression purchases.

Two instruments were used for flame atomic absorption analysis, a Perkin-Elmer Model 303 and a Perkin-Elmer Model 306 Atomic Absorption Spectrophotometer. Both instruments use a double-beam optical system. Slit width, burner head, flame settings, wavelength, and other instrumental settings are outlined in the Perkin-Elmer methods manual (10). To avoid errors which might arise from multiple dilutions of excessively high concentrations of certain elements, the burner head was turned 90°, giving a greatly shortened pathlength through the flame (a few millimeters instead of 10 cm) and thus lowering the sensitivity by an order of magnitude or more without making dilutions. Further decrease in sensitivity was required for sodium in some samples and was achieved by using the 330-nm resonance line, which is about 185 times less sensitive than the usual 589-nm line.

COLD VAPOR ATOMIC ABSORPTION

The most sensitive and reliable method for the determination of trace mercury levels in solution by atomic

TABLE V. - Atomic absorption parameters by element

Element	[1]λ, nm	[2]Flame	[3]Sens. µg/ml	Interferences	Interference correction
Ag	328.1	A	0.06	> 5% H_2SO_4 depresses	Dilution
Al	309.3	C	1.0	Ionization, > 0.2% Fe, HCl or H_2SO_4, depress	1,000 µg/ml Na, dilution
As	193.7	A	0.8	High salt content, false signal	D_2 and non-resonance line background correction
Ba	553.6	C	0.4	Ionization	1,000 µg/ml Na
Be	234.9	C	0.03	> 500 µg/ml, Al, Si, Mg, depress	Standard addition or add oxine
Bi	223.1	A	0.4	--	--
Ca	422.7	B	0.08	Ionization, Si, Al, $SO_4^=$, depress	1,000 µg/ml Na
Cd	228.8	A	0.03	--	--
Co	240.7	A	0.15	--	--
Cr	357.9	B	0.1	Fe and Ni, depress	Add 2% NH_4Cl
Cu	324.7	A	0.09	--	--
Fe	248.3	A	0.12	HNO_3, Ni depress	Very lean hot flame
Hg	253.6	Cold vapor only	[4]0.0002	SO_2, organic vapor, give false signal	Acid digestion and aeration
K	766.5	A	0.04	Ionization	1,000 µg/ml Na
Li	670.8	A	0.04	--	--
Mg	285.2	A	0.007	Si and Al depress	1,000 µg/ml Na
Mn	279.5	A	0.06	--	--

TABLE V. - Continued

Element	$^1\lambda$, nm	^{2}Flame	^{3}Sens. µg/ml	Interferences	Interference correction
Na	589.	A	0.02	Ionization	1,000 µg/ml KCl
Ni	232.0	A	0.15	--	--
Pb	283.3	A	0.5	--	--
Sb	231.2	A	1.0	--	--
Se	196.0	A	0.5	--	--
Sn	224.6	B	2.4	--	--
Sr	460.7	B	0.1	Ionization, Si, Al, P, depress	1,000 µg/ml Na
Tl	276.8	A	0.5	--	--
Zn	213.9	A	0.02	--	--

^{1}Wavelengths listed are the most sensitive for that element. Less sensitive wavelengths were used to avoid dilutions when high concentrations of elements were encountered, e.g., Na - 330 nm

^{2}Flames were as follows:
  A = Oxidizing (lean, blue) air - C_2H_2
  B = Reducing (rich, yellow) air - C_2H_2
  C = Reducing (rich, red) N_2O - C_2H_2

^{3}Sensitivity is approximated for 1% absorption as given in 10. Detection limits are dependent on the condition of the lamp, instrumental parameters, solution composition, etc., and are rarely as good as the sensitivity might indicate

^{4}The Hg cold vapor method used has a sensitivity of about 0.01 µg. The 0.0002 µg/ml limit is for a 50 ml sample

absorption is the "cold vapor" method, first described by
Hatch and Ott (11) and used for our work as modified by
Perkin-Elmer (Figure 3).

   Because the possibility of broad-band absorption of
the 253.7-nm Hg resonance line by organic or other fumes
is quite high, a few milliliters of 1:1 HCl were added to
each sample, and the sample was aerated into the absorption
cell of the system just prior to the addition of stannous
chloride to check for false absorption readings. If any
were detected, the system was flushed completely of such
interferences prior to mercury reduction by the stannous
chloride reagent. False readings from condensed water
vapor in the absorption cell were prevented by heating the
entire absorption cell area with a 250-watt infrared heat
lamp. This proved much more satisfactory than the
magnesium perchlorate drying cell initially provided with
the apparatus. The chemical removal of moisture constantly
varied the flow rates, and the magnesium perchlorate
rapidly became saturated as water vapor was absorbed during
a run.

   After running a sample, the closed-cycle air flow was
directed through the activated carbon scrubber (see
Figure 3), and all mercury was removed from the sample.

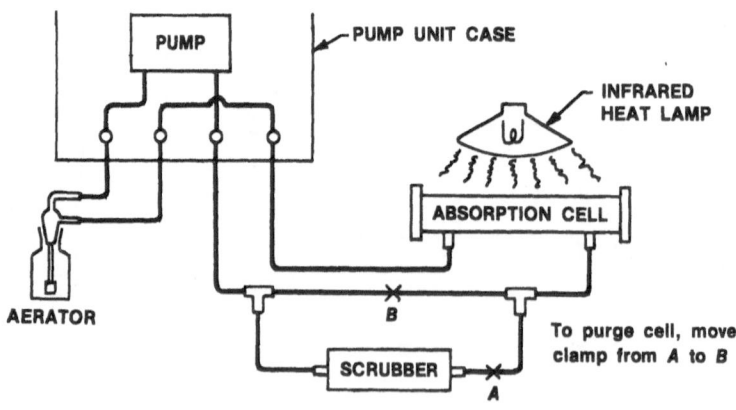

Figure 3. - Cold vapor mercury analysis system

No mercury vapors are vented into the laboratory using
this system. Passing fumes from an ICl solution through
the activated carbon scrubber was found to increase the
rapidity, effectiveness, and mercury collection lifetime of
the activated carbon severalfold.

The sample solution, now purged of all original
mercury, was then used as the matrix for standards. An
appropriate volume (usually 1 ml or less) of a standard
mercury chloride solution was pipetted into the aerator
flask, which still contained the previous stannous
chloride; the flask was stoppered immediately, and aeration
started, resulting in a standard peak. A standard curve
was made in this way, purging the solution of all mercury
between each standard and insuring that standards bracketed
the reading for each unknown. This procedure was necessary
because of the varying matrices of the samples.

FLAMELESS ATOMIC ABSORPTION

A second nonflame technique for atomic absorption
involves the use of a heated graphite furnace for atomization in place of the conventional flame. There are no
advantages in using the graphite furnace for atomic
absorption analysis where acceptable flame techniques are
available. The precision and accuracy of flame results can
be as much as an order of magnitude better than that of
flameless results, and the flameless methods require an
order of magnitude more instrument time (12). However,
detection limits can be lowered one or perhaps two orders
of magnitude by using the graphite furnace flameless atomic
absorption. It was used in this study for important
elements too low in concentration to be detected by
conventional flame techniques.

A Perkin-Elmer HGA 2100 Graphite Furnace was used with
the Model 306 spectrophotometer. A Perkin-Elmer Deuterium
Background Corrector was used in every determination, and
such correction is essential to minimize false signals from
absorption by smoke, water vapor, organic vapors, etc.

The Perkin-Elmer manual for the HGA 2100 (13) was used
as a guide for furnace temperatures and times, but
variations of these parameters can be found in other

Perkin-Elmer literature and in publications. Table VI lists
the various parameters we found most reliable for the
incinerator aqueous effluents and dissolved solids.

A standard addition technique was used as a check for
each element where matrix interferences were suspected.
Also, because standard addition will not correct for background and because the deuterium lamp does not completely
compensate for background absorbances much above 0.5
absorbance, a suitable spectral line close to the resonance
wavelength of the element of interest was used to check for
false absorption. For example, a total salt content above 1
weight-percent will produce an apparent absorption at the
193.7-nm As line, even when the metal is absent. The
non-ground-state mercury line at 194.2 nm was used to
determine the magnitude of this background.

Matrix modification was used for elements too volatile
to allow complete charring of interfering materials. For
example, 25 µl of 1000-ppm Ni was introduced into the
furnace along with every 20-µl aliquot of sample and
standard being analyzed for arsenic. During the heating of
the graphite furnace, a higher boiling nickel arsenide is
formed, thereby allowing the charring temperature to be
raised from 300° C to 1,200° C without loss of As.
Essentially all interfering materials are burned off at this
higher temperature, whereas the determination of As using
the lower charring temperature was impossible for most of
the samples in this study because interfering absorbances
were higher than could be compensated for by the D_2 background corrector.

UNDISSOLVED SOLIDS BY ATOMIC ABSORPTION

After being dried and powdered, the solids filtered
from the various aqueous samples were prepared for atomic
absorption analysis by the following acid digestion
method (14):

1. Accurately weigh a 1- to 5-gram sample. Transfer
    to a 250-ml Pt dish.

2. Add 50 ml of conc. HNO_3, or enough to cover the
    sample.

TABLE VI. - [1]Graphite furnace operating parameters

Element	λ, nm	Charring Temp, °C	Charring Time, s	Atomization Temp, °C	Atomization Time, s	[2]Sensitivity, picograms
Ag	328.1	400	40	2,600	8	3
Al	309.3	1,400	10	2,700	4	70
As	193.7	[3]1,200	40	2,700	3	20
Cd	228.8	250	30	2,100	5	1.5
Cr	357.9	1,100	30	2,700	4	20
Cu	324.7	900	22	2,700	4	30
Mn	279.5	1,100	20	2,700	4	4
Ni	232.0	1,000	20	2,700	4	100
Pb	283.3	625	30	2,700	4	20
Sb	217.6	900	40	2,700	3	25

[1]All elements were run using the same argon purge gas flow in the flow interrupt mode, with a drying temperature of 110° C and using the deuterium lamp background corrector. Drying times varied, depending on the sample size, e.g., 10 μl - 15 sec, 20 μl - 20 sec, 50 μl - 40 sec (13).

[2]Sensitivity is picograms giving an absorbance of 0.004 as listed in the manual (13).

[3]The 1,200° C charring temperature was made possible by the addition of 1,000 ppm Ni to each aliquot.

3. Digest slowly over low heat for several hours with
    a Pt cover on the Pt dish until much of the organic
    has decomposed. Replenish with HNO_3 as needed.

4. Add 10 ml of 48 percent HF. Continue digestion
    under Pt cover over low heat, frequently rinsing
    the sides of the Pt dish with distilled water to
    assure that all material reacts with the acid
    mixtures and to prevent loss of salts forming on
    the edge of the Pt dish. Digestion should continue
    until the solution becomes clear, and relatively
    free of black particles. Concentrated HNO_3 is
    added as needed. Two or three days were often
    required.

5. Transfer to the perchlorate hood. Add 10 ml of
    $HClO_4$. Continue digestion under cover over low
    heat with frequent rinsing of the Pt dish walls
    until all carbonaceous material has been
    destroyed. <u>An excessively high hotplate
    temperature may cause an explosion if carbonaceous
    material is present</u>. About 20 ml more of HF is
    added during this digestion.

6. Remove the cover, elevate the temperature,
    evaporate until fumes of perchloric acid appear.
    Cool.

7. Dissolve the solids in a solution of 5 ml HCl and
    50 ml water using low temperature.

8. Transfer to a 100-ml volumetric flask and dilute to
    volume with 1-percent-HCl solution.

Many other dissolution approaches were tried, including
ashing prior to acid digestion, using sulfuric acid as well
as nitric acid, omitting the perchloric acid, and fusing the
remaining undissolved solids with potassium pyrosulfate or
sodium carbonate. The ashing, even at 500° C, was suspected
of driving off volatiles such as arsenic. Sulfuric acid
interferes in the atomic absorption analysis of several
elements (see table V) and results in sulfate precipitates
(for example, Ba, Pb). The fusion techniques result in the
addition of Na or K, and separate samples had to be prepared
to determine these elements. Consequently, the above-

outlined acid digestion method was used, and resulted in
essentially complete dissolution of the solids if allowed
to digest slowly. At times, a small amount of the original
sample remained undissolved (<1 percent), but emission
spectrographic analysis revealed that these solids consisted
primarily of Si, indicating the HF digestion had been
incomplete.

To determine the fate of mercury during this acid
digestion process, radioactive mercury tracer was added
to several samples. The radioactive mercury tracer
remaining in the final analytical solution ranged from 13
to 48 percent of the total amount added. Therefore, to
determine mercury in the solids using this acid digestion
procedure, it was necessary to use radioactive mercury as a
collection monitor.

As a further check of the acid digestion atomic
absorption method, portions of the NBS Standard Reference
Material 1633 Coal Fly Ash were analyzed periodically along
with the solid unknowns. Results are compared in table VII
with the NBS certified values and with the average obtained
by four independent laboratories using instrumental neutron
activation analysis (INAA) for most determinations (15).
Where discrepancies are apparent, for example, Cd, Na, and
Tl, new standards were prepared from primary materials,
standard addition was used, and an analysis from another
analytical chemistry laboratory was obtained. These
discrepancies still remain and are currently under
investigation.

Reagent blanks, of course, can be a problem for several
elements using the quantities of acids required for this
procedure. Using reagent-grade chemicals, the highest
reagent blank concentrations found were around 3 ppm for
Ca, Fe, and Na. Reagent blanks should be run with each set
of unknowns in the acid digestion atomic absorption
procedure.

An idea of the concentration levels of the elements in
the fly ash from the incinerator, and the deviation of
elemental composition from sample to sample along with
analytical variation, can be seen in table VIII. The
samples analyzed by the two different methods were not the
same, nor were they collected during the same time period.

TABLE VII. — Comparison of analyses of SRM 1633 coal fly ash

Element	Concentration, ppm		
	Acid Digestion-AA	NBS[1]	INAA[2]
Ag	3 ± 1	—	—
Al	118,000 ± 5,000	—	127,000 ±5,000
As	76 ± 14	61 ± 6	58 ± 4
Ba	2,500 ± 400	—	2,700 ± 200
Be	16 ± 1	[3](12)	—
Bi	61 ± 14	—	—
Ca	45,500 ± 500	—	47,000 ±6,000
Cd	6.0 ± 1.6	1.45± .06	—
Co	42 ± 2	[3](38)	41.5± 1.2
Cr	130 ± 4	131 ± 2	127 ± 6
Cu	128 ± 14	128 ± 5	—
Fe	64,000 ±2,000	—	62,000 ±3,000
Hg	.19± .05	.14± .01	—
K	16,800 ±1,900	[3](17,200)	16,100 ±1,500
Li	62 ± 24	—	—
Mg	17,800 ± 800	—	18,000 ±4,000
Mn	490 ± 20	493 ± 7	496 ± 19
Na	5,600 ± 800	—	3,200 ± 400
Ni	100 ± 4	98 ± 3	98 ± 9
Pb	82 ± 9	70 ± 4	75 ± 5
Sb	[4]<30	—	6.9± .6
Se	[4]<15	9.4± .5	10.2± 1.4
Sn	270 ± 40	—	—
Sr	1,200 ± 200	[3](1,380)	1,700 ± 300
Ti	7,100 ± 500	—	7,400 ± 300
Tl	46 ± 2	[3](4)	—
Zn	212 ± 9	210 ±20	216 ± 25

TABLE VII. - <u>Comparison of analyses of SRM 1633 coal fly ash</u> - Continued

[1]Certified by the National Bureau of Standards to be "...in no case less than the 95% confidence limits computed for the analysis."

[2]Instrumental Neutron Activation Analyses reported by Ondov, et al., 7th Materials Research Symposium, P. D. LaFleur, ed., NBS (1975) (15).

[3]Values in parentheses ( ) are noncertified values included by NBS for information only.

[4]Background absorbances too high to use the graphite furnace atomic absorption for lower detection limits.

Yet the agreement for some elemental concentrations shows surprisingly little variation between the two sampling periods.

TABLE VIII. - <u>Elemental composition of fly ash from scrubbers</u>

Element	Concentration units	INAA samples	AA samples
Ag	ppm	[1]106 ± 18	[1]130 ± 30
Al	%	11.1 ± 1.2	12.4± 1.2
Ca	%	4.0 ± 0.6	2.3± 0.8
Cd	ppm	45. ± 28	64 ± 16
Co	ppm	33 ± 4	100 ± 30
Cr	ppm	1,430 ±200	1,200 ±530
Cu	ppm	1,000 ±500	510 ±150
Fe	%	4.6 ± 0.7	2.4± 0.8
Mn	ppm	3,400 ±700	1,500 ±700
Na	%	1.4 ± 0.2	1.6± 0.2
Sb	ppm	260 ± 60	[2]330 ±300
Zn	%	0.94± 0.13	1.1± 0.2

[1]Uncertainties represent the standard deviations of samples taken at different periods of time and are mainly the result of sample-to-sample variations rather than analytical error.

[2]Only three samples run for Sb by flameless AA: 165 ppm, 170 ppm, and 665 ppm.

TREATMENT OF DATA

Before effluents from an incinerator or any other
source can be identified as a source of pollution for
certain elements, it is necessary for meaningful comparisons to be made. For example, quench water effluents from
the Alexandria incinerator contain up to 4.7 ppm Mg, but
the input tap water contains up to 4.2 ppm. Actually,
most of the quench waters were found to be lower than the
tap water in Mg so these waters can hardly be considered a
source of Mg. Enrichment of an element above its tapwater
concentration can therefore provide meaningful information
concerning the source of wastewater pollutants.

To determine whether particles in the air are simply
suspended crustal material or are from a manmade source,
an enrichment factor, EF, has been defined (16):

$$EF = \frac{(C_x/C_{Al})i}{(C_x/C_{Al})crust}$$

Fluctuations of absolute concentration are removed in this
definition by using a double normalization in which the
C's represent the concentrations of element X and Al in
medium i (in this case, atmospheric particles) and in the
average crustal material of the earth (17). For average
suspended crustal material, the EF's for all elements will
be close to unity. To identify the origin of enriched
elements with EF's much greater than unity, sources must
be found emitting particles of equal or greater EF values
for the same elements. Figure 4 gives a comparison of EF
values for several elements in particles from the
Alexandria incinerator, coal-fired powerplants, and the
ambient atmosphere of urban areas.

In considering the enrichment of elements on particles
generated by a certain process, it is important to know the
size of the particles concentrating the elements. It is
environmentally less desirable for toxic elements to
preferentially condense on the small particles because they
are less efficiently collected by pollution control devices
and are more efficiently deposited in human lungs. For
this reason cascade impactor studies are important, and
Figure 5 shows the results of such a study for Al and Zn
from the Alexandria incinerator.

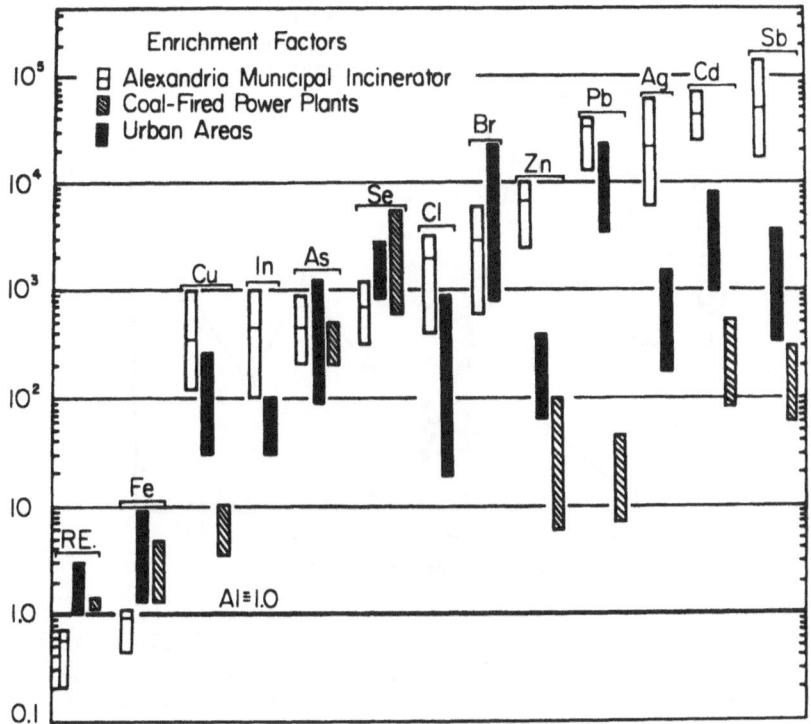

Figure 4. - Ranges of enrichment factors for several
elements on particles from six U.S. cities,
several coal-fired powerplants, and the
Alexandria (Va.) Municipal Incinerator (18).
The centerline in the bars for the municipal
incinerator represents the average value.

SUMMARY

Incinerator studies and similar studies of powerplants
and industrial plants are important for public health and
total environmental quality considerations, and the
analytical results need to be reliable. The analysis of
incinerator effluents is being performed in separate
laboratories by two different analytical methods:

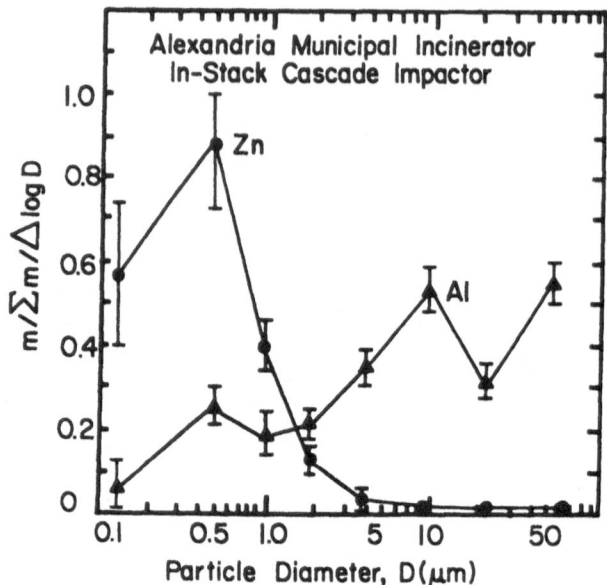

Figure 5. - Relative concentrations of particulate Al and
Zn in stack gases from the Alexandria incinerator as a function of particle diameter.
The fractional mass of each cascade impactor
stage (m/Σm) is normalized to the aerosol
effective cutoff diameter of each stage
(Δ log D). In placing points for the smallest
and largest diameters, it is assumed that most
of the particulate mass falls between diameters
of 0.05 and 50 μm.

instrumental activation analysis and atomic absorption
analysis. These two methods provide a complementary and
comprehensive coverage of most elements of environmental
interest in aqueous and solid effluents. Where overlapping of the two techniques has provided a basis for
comparison, greater confidence has been gained in both
methods to provide the kind of reliability required for
environmental decisions. Hopefully, this study will

provide ideas and insights into some of the analytical considerations involved in characterizing effluents from a system such as the Alexandria incinerator.

REFERENCES

1. Carotti, A. A. and R. A. Smith, "Gaseous Emissions from Municipal Incinerators," EPA Report SW-18C (1974).

2. Robertson, C. A. M., "The Emission of HCl from Municipal Incinerators and Related Studies," Solid Wastes Management (England), 64, 139 (1974).

3. Tanas, M. J., A. R. Greenberg, R. D. Hoak, and M. C. Rand, editors, Standard Methods for the Examination of Water and Wastewater, 13th Ed. (1971), Am. Pub. Health Ass., Wash., D.C., p. 417.

4. Methods for Chemical Analysis of Water and Wastes, EPA-625-/6-74-003, (1974), U.S. Office of Tech. Transfer, Wash., D.C.

5. Pilat, M. J., D. S. Ensor, J. C. Bosch, "Source Test Cascade Impactor," Atmos. Environ. 4, 671 (1970).

6. Pilat, M. J., "Measurement of Particle Size Distributions at Emission Sources with Cascade Impactor," Joint US-USSR Symposium on Control of Fine Particulate Emission From Industrial Sources, Jan. 14-18, 1974, San Francisco, Calif. M. J. Pilat, Assoc. Prof., Dept. of Civil Engineering, University of Washington, Seattle, Wash.

7. Martin, R. M., "Construction Details of Isokinetic Source-Sampling Equipment," Air Pollution Control Office, EPA, Research Triangle Park, N.C., Pub. #APTD-0581 (1971).

8. Rom, J. J., "Maintenance, Calibration, and Operation of Isolinetic Source-Sampling Equipment," Office of Air Programs, EPA, Research Triangle Park, N.C., Pub. #APTD-0576 (1972).

9. Harrison, S. H., P. D. LaFleur, and W. H. Zoller, "An
    Evaluation of Lyophilization for the Preconcentration
    of Natural Water Samples Prior to Neutron Activation
    Analysis," Anal. Chem. 47, (1975).

10. Perkin Elmer, Analytical Methods for Atomic Absorption
    Spectrophotometry, 1973, The Perkin-Elmer Corp.,
    Norwalk, Conn.

11. Hatch, W. R. and W. L. Ott, "Determination of Sub-
    Microgram Quantities of Mercury by Atomic Absorption
    Spectrophotometry," Anal. Chem. 40, 2085 (1968).

12. Morgenthaler, L., "A Primer for Flameless Atomization,"
    Amer. Lab., Apr. 1974, p. 41.

13. Perkin-Elmer, Analytical Methods for Atomic Absorption
    Spectroscopy Using the HGA Graphite Furnace, 1973,
    The Perkin-Elmer Corp., Norwalk, Conn.

14. Epstein, M. S., T. C. Rains, and O. Menis, "Determina-
    tion of Cadmium and Zinc in Standard Reference
    Materials by Atomic Fluorescence Spectrometry with
    Automatic Scatter Correction," Canadian J. of Spec.
    20, 22 (1975).

15. Ondov, J. M., W. H. Zoller, I. Olmez, N. K. Aras,
    G. E. Gordon, L. A. Rancitelli, K. H. Abel, R. H.
    Filby, K. R. Shah, and R. C. Ragaini, "Elemental
    Concentrations in the NBS Environmental Coal and
    Fly Ash Standard Reference Materials," Anal. Chem.
    47, 1102 (1975).

16. Zoller, W. H., E. S. Gladney, and R. A. Duce,
    "Atmospheric Concentrations and Sources of Trace
    Metals at the South Pole," Science 183, 198 (1974).

17. Wedepohl, K. H., Origin and Distribution of the
    Elements, L. H. Ahrens, Ed., Pergamon Press, London,
    (1968) pp. 999-1016.

18. Greenberg, R. R., W. H. Zoller, and G. E. Gordon,
    "Municipal Incinerators: Source of Toxic Elements on
    Urban Aerosols," submitted to Science for publication.

TRACE ELEMENT LEVELS IN SOUTHERN NEW JERSEY DRINKING WATERS

S. A. Katz and D. M. Scheiner

Rutgers University

Camden, N.J. 08102

The New Jersey State Department of Environmental
Protection has established potable water standards for use
in the administration of the laws and regulations relating
to water used for drinking and/or cooking purposes within
the State (1). These standards require the owner of a
public supply assure that the water is periodically tested
for bacteriological quality and make the results of these
tests available to the New Jersey Department of
Environmental Protection, the New Jersey Department of
Health or the local board of health. Although the potable
water standards include a section on chemical quality, no
direct statements are made regarding responsibility for
determining chemical quality, the frequency of chemical
evaluations or the agency to which the results of the
chemical quality evaluations are to be reported. Considering the lack of a direct statement in the standards, the
large number of municipalities in the State and the
complexities of local, county and state authorities, we
initiated a survey of the chemical quality of potable water
in southern New Jersey.

Southern New Jersey is roughly defined as that portion
of the State below Trenton consisting of Atlantic,
Burlington, Camden, Cape May, Cumberland, Gloucester, Ocean
and Salem Counties. Southern New Jersey encompasses
approximately 50% of the area of the State, and the
population of 1,750,000 is slightly less than 25% of the
State's total population.

Members of the Rutgers faculty, staff and student body

residing in southern New Jersey provided samples of their
domestic water for laboratory evaluation. Each month,
number-coded, 4 ounce, plastic sample bottles previously
cleaned with dilute hydrochloric acid and rinsed with
deionized water were distributed to those participating in
the survey. An instruction sheet advising the participants
to run water from the tap for several minutes and to rinse
the bottle several times before collecting the sample was
attached to each bottle. Also attached to each bottle was
a number-coded data sheet on which information regarding
the sample was provided by the participant.

Monthly samples from 53 southern New Jersey communities
were collected. In many instances, samples were obtained
from several domestic sites in each community. According
to the information provided by the participants and received
from the various water commissions, the samples came from
wells ranging in depth from 50 feet to over 500 feet. No
surface water samples were identified. The communities
from which samples were collected had populations ranging
from less than 600 to over 100,000. Of the samples
collected, 88% came from public water supplies; the remaining 12% were drawn from private domestic wells.

As each monthly set of samples was returned to the
laboratory, a 50 ml aliquot was removed from each bottle
and diluted to 100 ml with "TISAB" (total ionic strength
adjustment buffer) prior to measurement of fluoride ion
concentration by the electrode method. The water samples
remaining in the bottles were treated with two drops of
concentrated hydrochloric acid and retained for the analysis
of chromium, copper, iron, lead and zinc by atomic absorption spectrometry. Bottles cleaned with hydrochloric acid
and rinsed with deionized water were filled with deionized
water and treated as above to determine blank values.

The Corning fluoride sensitive electrode-saturated
calomel electrode pair was used in conjunction with a
Corning Model 12 expanded scale pH meter to determine the
fluoride ion concentrations of the samples. The procedures
followed were those of 121 B in "Standard Methods" (2).
The results of these measurements are summarized in figure
1.

Over 80% of the samples obtained from public water
supplies and over 90% of the samples obtained from private

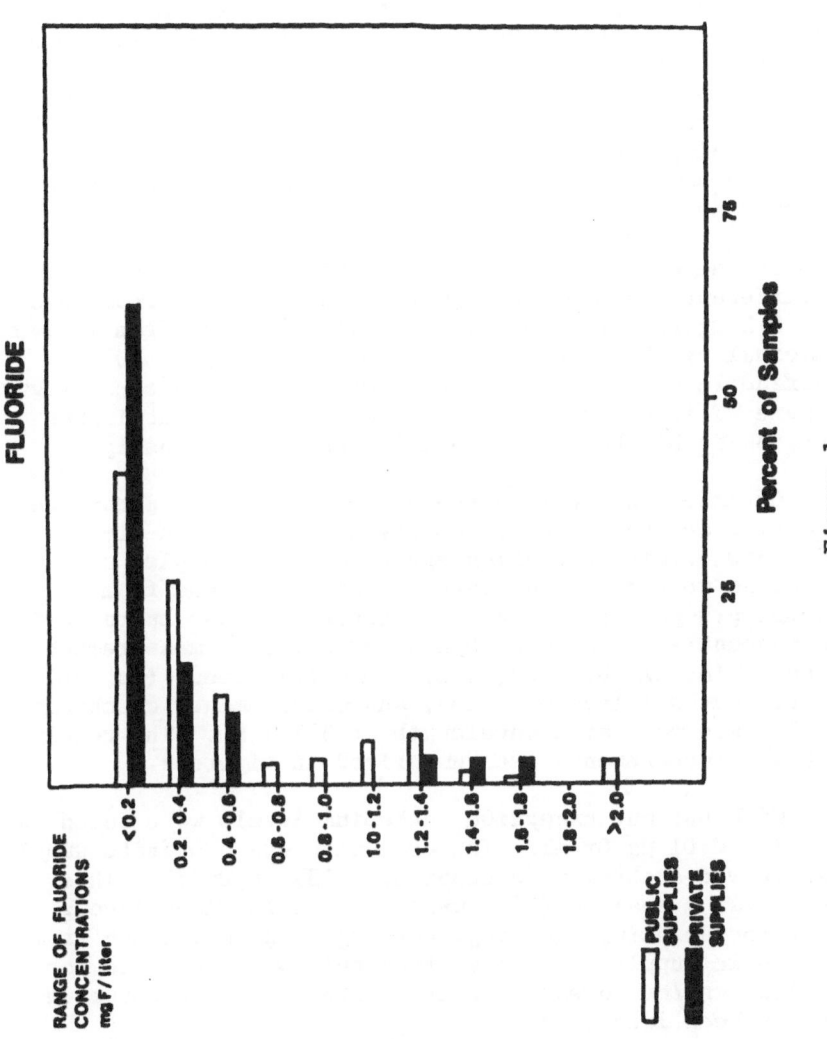

Figure 1

supplies showed fluoride levels of less than 0.7 µg F/ml.
Although controlled fluoridation of drinking water is rare
in southern New Jersey, this was the cause of fluoride
levels greater than 0.7 µg F/ml in Egg Harbor City,
Willingboro and possibly Burlington City. In the case of
Clarksboro, Glassboro, Mantua, Mullica Hill, Pitman, Sewell,
Wenonah, Woodbury and Woodstown, natural fluoride was
responsible for fluoride levels greater than 0.7 µg F/ml.
In these instances, water is drawn from deep (500' – 700')
wells tapping the Raritan stratum (3). Only in the
Woodstown samples, however, did the fluoride levels exceed
the recommended maximum of 2 µg F/ml.

In the course of our investigation, the fluoride levels
of different municipalities were found to range from less
than 0.2 µg F/ml to over 2.5 µg F/ml. Samples from a given
municipality, however, showed fairly constant (± 20%)
fluoride levels. Some select values are presented in figure
2 where WT is Woodstown, WB is Woodbury, B is Burlington
City, BW is Blackwood and CH is Cherry Hill Township.

Chromium, as well as the other metals, was determined
by atomic absorption spectrometry using a Perkin-Elmer
Model 360 atomic absorption spectrometer. Samples and
standards were aspirated into the air-acetylene flame
without pre-concentration. The instrument was operated in
the concentration mode with a lower limit of measurement
of approximately 0.01 µg Cr/ml. A multielement (Co, Cr,
Cu, Fe, Mn, Ni) lamp was used, and measurements of chromium
levels were made at a wavelength of 357.9 nm. The results
of these measurements are summarized in figure 3.

With but one exception, chromium levels were found to
be below 0.01 µg Cr/ml. The exception was a private supply
that showed a chromium content of 0.139 µg Cr/ml. This
supply was a shallow (30') well located in Cinnaminson.
Water for drinking, cooking, bathing, etc. was drawn from
the public supply, and water from this well was used for
the lawn and/or to wash the car. The source of chromium
has not been identified.

Copper levels were determined by essentially the same
procedures used for chromium levels with the exception that
the measurements were made at a wavelength of 324.7 nm
under oxidizing flame conditions. Scale expansion allowed
a limit of measurement of 0.01 µg Cu/ml. A summary of the

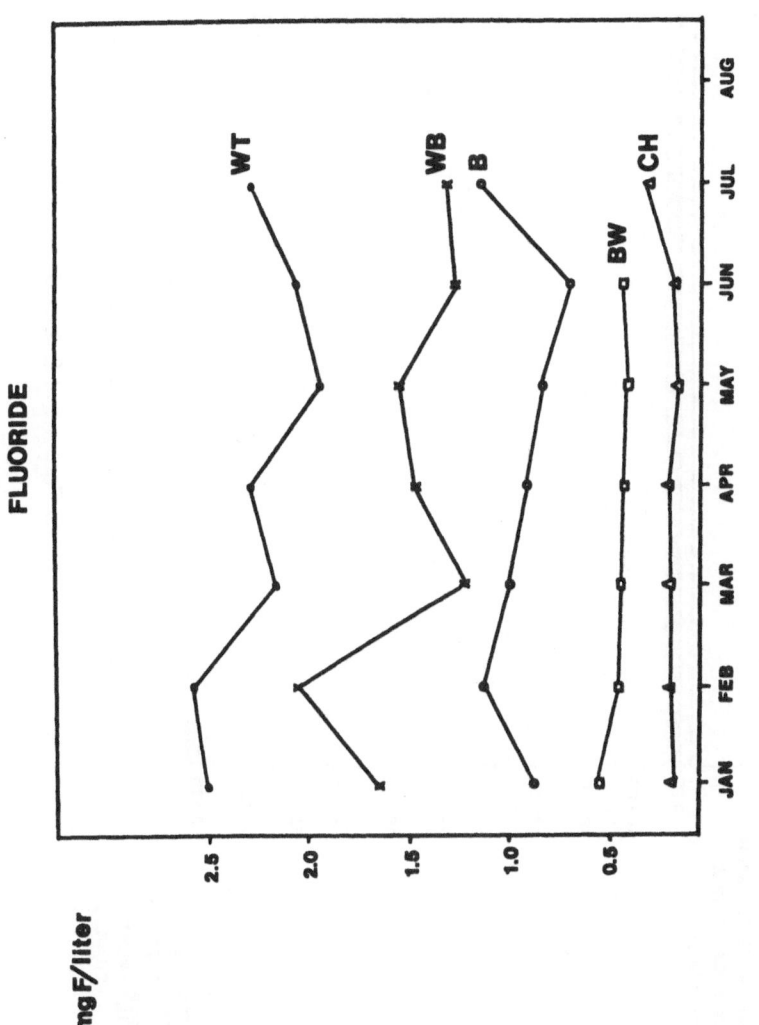

Figure 2

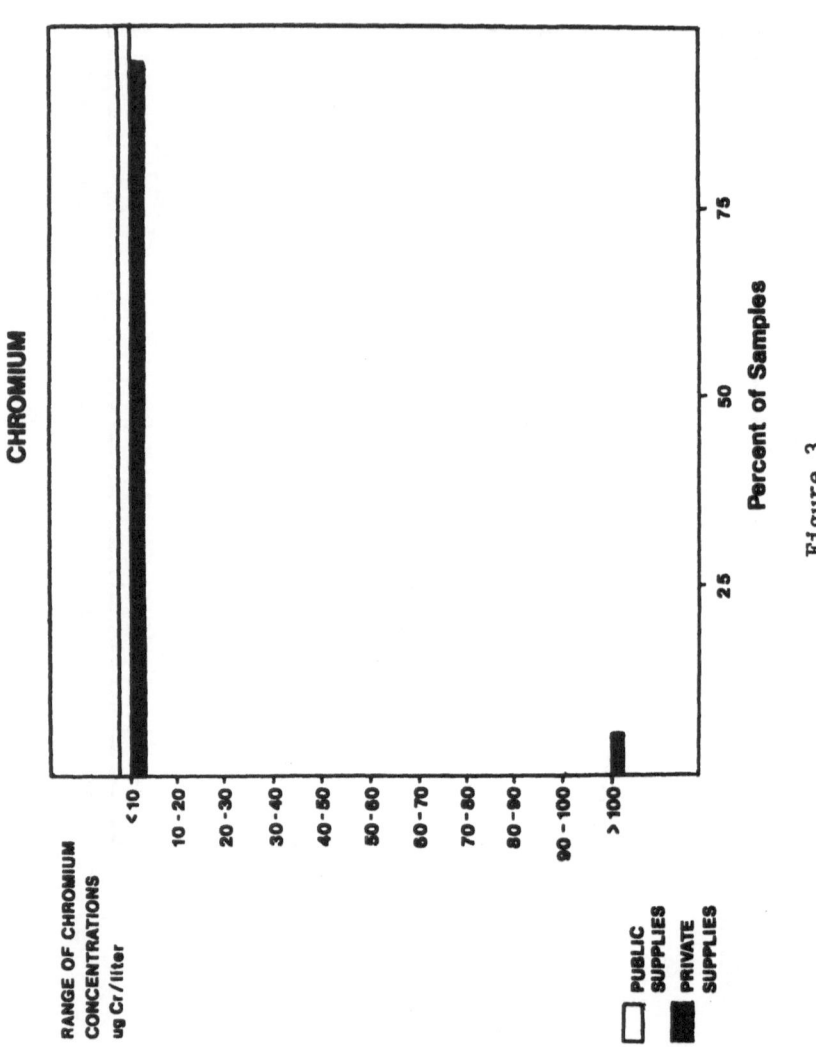

Figure 3

results of these measurements is presented in figure 4.

Over 95% of the samples taken from public supplies and approximately 85% of the samples taken from private supplies had copper levels below 0.5 μg Cu/ml, well below the recommended maximum of 1 μg Cu/ml (1). The samples taken from public supplies showing copper levels above 1 μg Cu/ml were inconsistent on a month-to-month basis. In one instance, the high copper levels were traced to an acid cleaning of the coils in a summer-winter hot water system. Samples from other locations served by the same public supply showed consistently lower copper levels. Public supplies do, however, show significant variations in copper levels on a month-by-month basis. This is reflected in figure 5 where copper levels of potable water samples taken from Maple Shade Township, Haddon Township and Cherry Hill Township are presented.

Some 15% of the samples taken from private supplies showed copper levels above 1 μg Cu/ml. Most of these samples came from private domestic wells located in Cinnaminson, Collins Lakes and Erial. Twelve additional samples were collected over a three week period from the Erial well. Copper levels of these twelve samples ranged from 1.25 to 4.30 μg Cu/ml. The source of copper has not been identified.

Iron levels were also measured by atomic absorption spectrometry. Measurements were made at a wavelength of 248.3 nm. Operating the instrument in the concentration made with scale expansion allowed a lower measurement limit of 0.01 μg Fe/ml. The results of these measurements are summarized in figure 6.

Of the samples evaluated, 26% of those from public supplies and 75% of those from private supplies exceeded the recommended maximum iron level of 0.3 μg Fe/ml (1). Samples from many of the public supplies showed variations of two fold on a month-to-month basis; i.e., the extremes of the ranges of iron levels differed by approximately a factor of two. Samples from several sites within the same public supply also varied from each other by factors of two or three during any given month in most cases. In some cases, variations as high as ten fold were observed in the extremes of the ranges of iron levels in public supplies. Potable water samples collected from two sites in Maple

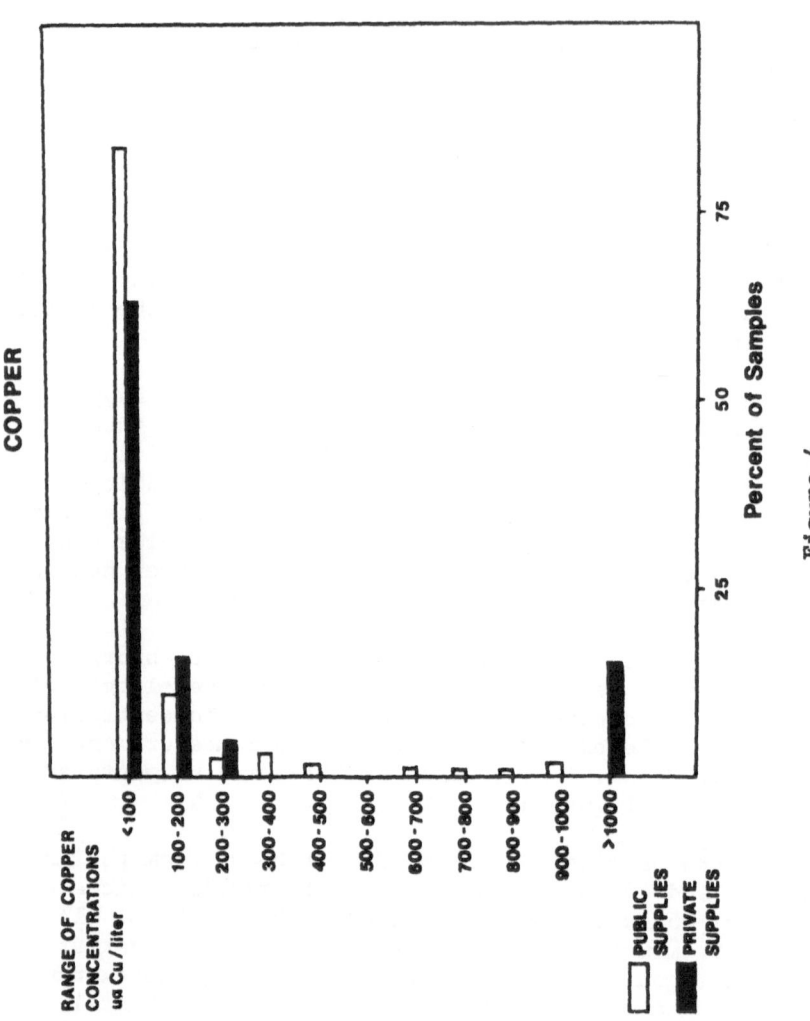

Figure 4

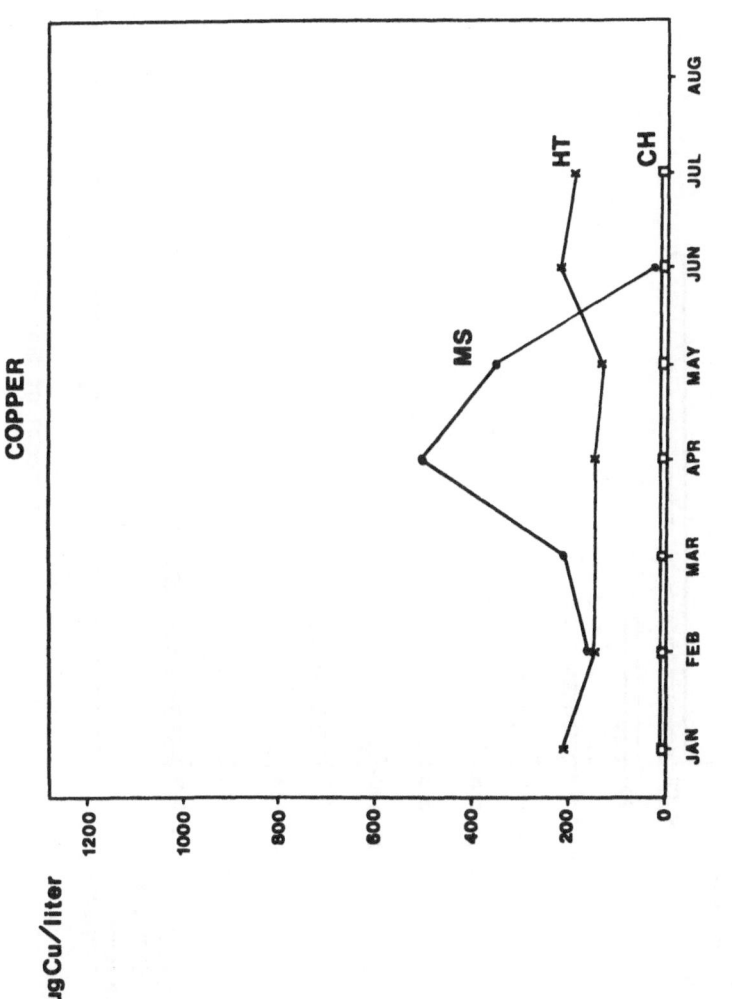

Figure 5

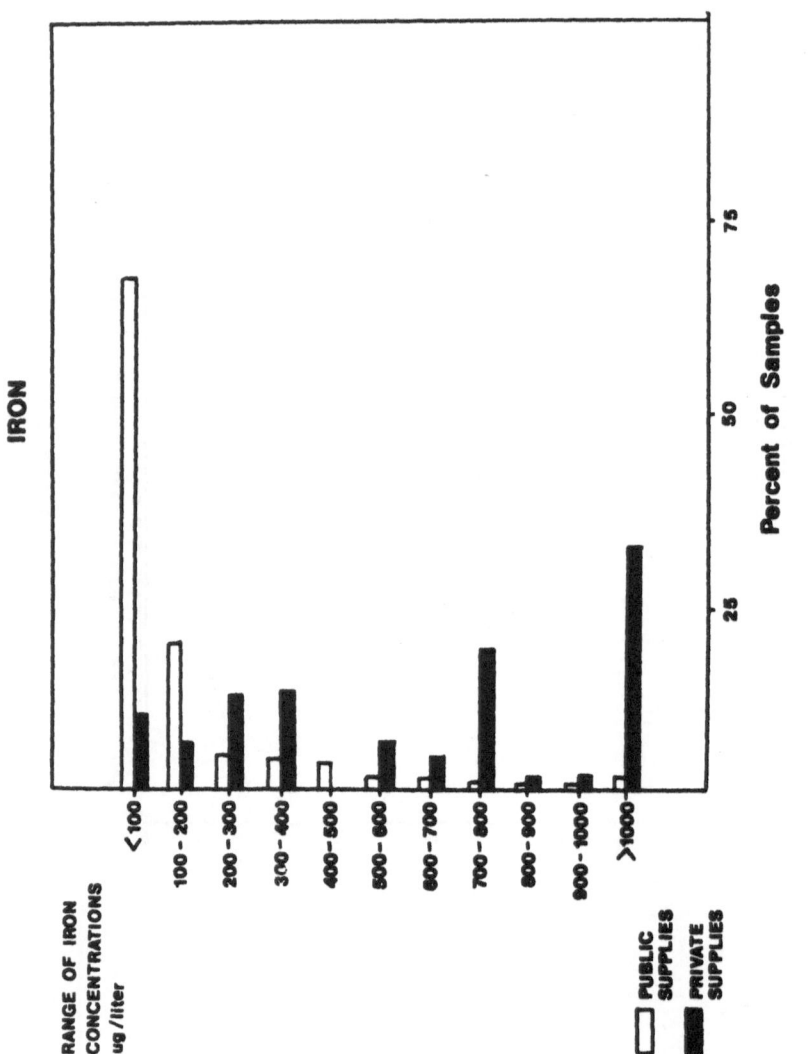

Figure 6

Shade Township demonstrated such a wide range of iron levels
as shown in figure 7. In this case, the water supply rather
than the domestic plumbing systems was probably responsible
for the observed iron levels. In other cases, the source
of iron could not be immediately identified.

The iron levels of samples collected from private
supplies were frequently much higher than those of public
supplies. Variations of from 2 to 4 fold were observed in
the ranges of iron levels in private supplies. While some
of the private supplies consistently showed iron levels
below 0.3 µg Fe/ml, most were in one of two categories;
those with iron levels ranging from 0.5 to 1 µg Fe/ml or
those with iron levels ranging from 1.5 to 7.5 µg Fe/ml.
The water supply was, in most cases, probably responsible
for the observed values.

Lead levels were measured at a wavelength of 283.3 nm
using a lead hollow cathode lamp and operating the
instrument in the concentration mode. With scale expansion,
the limit of detection was 0.025 µg Pb/ml. Of all the
samples examined, lead was identified in only two sources
both of which were public drinking fountains on the Rutgers
campus. Additional samples were collected from these
fountains and from the adjacent rest rooms. Lead was not
detected in the water samples from the rest rooms, but
values consistently in the 0.1 µg Pb/ml range were obtained
from the samples taken from the drinking fountains. It has
not yet been determined if the fountains or their
installations were responsible for the observed values.

A zinc-calcium hollow cathode lamp was used for the
determination of zinc levels in the potable water samples.
Measurements were made at a wavelength of 213.9 nm with the
instrument operating in the concentration mode. With scale
expansion, the lower limit of measurement was 0.01 µg
Zn/ml.

Of the samples collected from public water supplies,
90% had zinc levels of less than 0.1 µg Zn/ml, and none of
the samples exceeded 1 µg Zn/ml. The zinc levels of
samples from the same source showed variations of two to
three fold on a month-to-month basis, and samples taken
from several sites within the same public distribution
system during the same month varied by as much as a factor

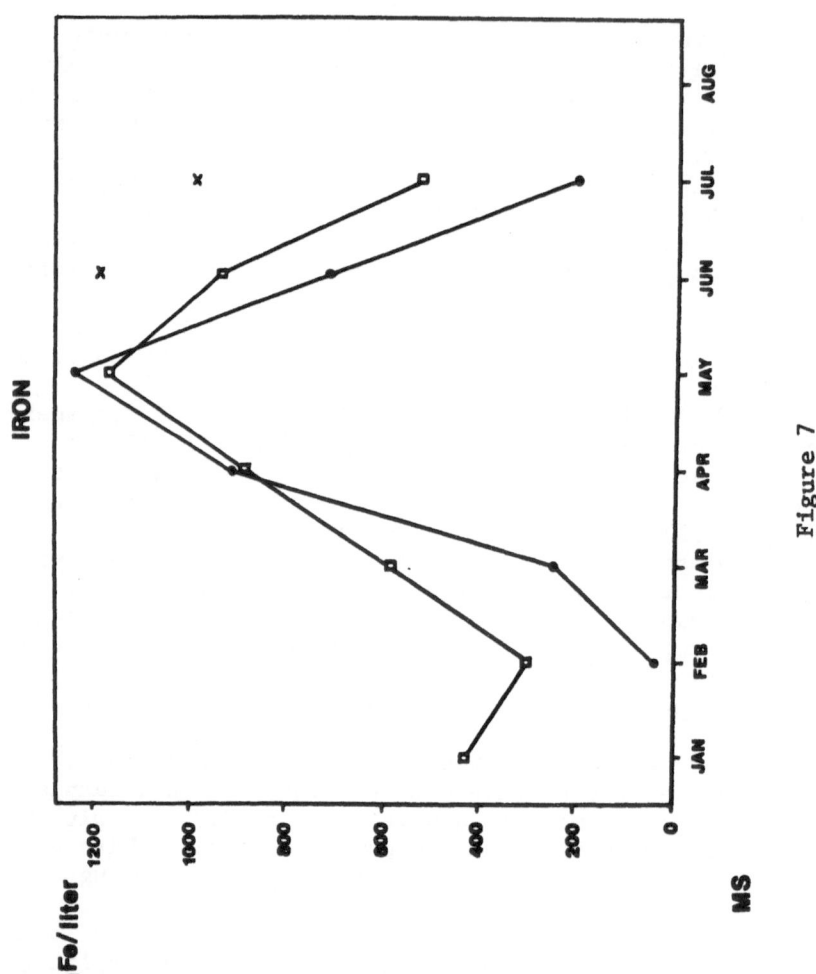

Figure 7

of three in some cases.

Zinc levels above 1 µg Zn/ml were observed in 28% of
the samples collected from private supplies. Only one
private supply, however, showed zinc levels above the
recommended maximum of 5 µg Zn/ml (1). This supply was a
private domestic well located in Monroeville, and zinc
levels ranging from 8.5 to 13 µg Zn/ml were observed. The
source of zinc in this supply has not been identified.

The foregoing data which are summarized in the table
were obtained from monthly samplings in some 50 southern
New Jersey communities. These samplings do not represent
all potable water supplies in the southern part of the
State, but some general comments regarding the data may be
in order.

In general, public water supplies appear to meet the
potable water requirements more frequently than do the
private supplies. This tendency is most pronounced with
respect to iron levels, but copper and zinc levels also
appear to be higher in water from private domestic wells
than in water from public supplies. On the basis of all
parameters evaluated for all of the samples collected, the
recommended levels were exceeded seven times more frequently in samples from private supplies than in those from
public supplies. The apparent inferiority of private
supplies become more significant in light of a report that
as late as 1963, more than one half of the communities in
New Jersey (representing approximately 14% of the State's
population) did not have public water supplies available
(4).

The variations in trace element levels of public
supplies on a month-to-month or site-to-site basis negates
the validity of water quality evaluations based on a single
sample. Chloride levels have been reported to double over
only a thirty minute period in Wildwood and in Atlantic
City (3), and the iron and manganese contents of water from
the same well are reported to show variations when
collected at different times (5). In the course of the
present study, copper levels in water from a private
domestic well were found to decrease from 1.4 to 0.1 µg
Cu/ml between 10:00 a.m. and noon. In view of the apparent
variations in trace element levels in public water supplies,

Trace Element Levels in Southern New Jersey Drinking Waters

Range of values observed, μg/ml

Location	F	Cr	Cu	Fe	Pb	Zn
Atco*	0.1	ND	ND	0.3	ND	4.4
Audubon	0.2	ND	0.5	< 0.1	ND	< 0.1
Avalon	0.2-0.3	ND	ND -0.1	ND -0.1	ND	0.1-0.5
Barrington	0.2	ND	0.2	< 0.1	ND	ND
Bellmawr	0.4-0.5	ND	ND -0.1	0.1-0.5	ND	< 0.1
Blackwood	0.4-0.5	ND	ND -0.2	0.1	ND	< 0.1
Burlington City	0.9-1.1	ND	ND -0.1	0.1	ND	< 0.1
Camden City	0.1-0.3	ND	ND -0.3	0.2-2.0	ND -0.5	0.1-0.9
Cherry Hill	0.1-0.3	ND	ND -0.3	< 0.1	ND	< 0.1
Cherry Hill*	0.1	ND	ND -0.2	0.7-0.9	ND	0.1
Cinnaminson	0.1	ND	ND -0.1	ND -0.1	ND	< 0.1
Cinnaminson*	0.1	0.1	0.1-1.4	1.6-6.4	ND	1.0-3.0
Clarksboro	1.1-1.4	ND	ND	0.5-0.9	ND	< 0.1
Clementon	0.1-0.2	ND	ND -0.1	0.1-0.3	ND	0.1
Collingswood	0.1-0.2	ND	ND	0.1-0.2	ND	0.1
Collins Lakes*	0.1	ND	0.2-4.6	2.6-4.6	ND	0.1
Edgewater Park	0.1	ND	ND	0.1	ND	< 0.1
Egg Harbor City	0.9-1.3	ND	ND	ND -0.1	ND	0.1
Erial*	ND	ND	1.0-4.2	0.2-0.7	ND	0.1
Forked River*	0.1-0.2	ND	ND	0.6-0.7	ND	2.0
Gibbsboro*	0.1-0.2	ND	ND	0.4-0.5	ND	0.1-0.4
Glendora	0.5	ND	ND	0.1	ND	< 0.1
Gloucester City	0.1-0.2	ND	ND -0.1	0.1-1.5	ND	ND

Haddonfield	0.1-0.2	ND	ND	ND -0.3	ND	ND -0.1
Haddon Township	0.1-0.2	ND	0.1-0.2	ND -0.1	ND	ND
Haddon Heights	0.9	ND	ND	< 0.1	ND	ND
Hi-Nella	0.2-0.3	ND	0.1-0.2	ND -0.1	ND	0.1-0.3
Lindenwold	0.3-0.5	ND	ND -0.1	ND -0.1	ND	ND
Mantua*	1.1-1.6	ND	0.1-0.5	0.2	ND	0.3
Maple Shade	0.2-0.3	ND	ND -0.1	0.2-2.4	ND	ND -0.1
Marlton	0.1-0.2	ND	ND	0.1-0.2	ND	ND
Marlton*	0.3	ND	ND	0.2	ND	ND
Medford	0.1	ND	ND	0.1	ND	ND
Medford*	0.2-0.3	ND	ND	0.3-1.4	ND	ND -0.8
Medford Lakes*	0.5	ND	ND	0.1-0.9	ND	0.2-0.8
Merchantville	0.1	ND	ND	< 0.1	ND	ND
Monroeville*	0.1	ND	ND -0.1	4.7-7.4	ND	9.0-13
Moorestown	0.1-0.2	ND	ND	0.1	ND	ND -0.1
Mullica Hill*	0.4	ND	ND	2.5-3.9	ND	0.7
Oaklyn	0.2-0.3	ND	0.1-0.7	0.1-0.6	ND	ND -0.1
Ocean City	0.2-0.4	ND	ND	ND	ND	0.1
Palmyra	0.1	ND	ND -0.6	ND -0.1	ND	ND
Pennsauken	0.1-0.2	ND	ND -0.2	ND -0.3	ND	ND
Pennsville	0.1-0.6	ND	ND	ND -0.1	ND	ND
Pine Hill	0.4	ND	0.1	0.1	ND	ND
Somerdale	0.3-0.5	ND	0.1	ND	ND	ND -0.7
Stone Harbor	0.2-0.3	ND	ND	ND -0.3	ND	< 0.1
Sweetwater*	0.1	ND	ND	1.0-3.2	ND	1.2-3.4
Vincentown	0.1	ND	ND	0.1	ND	ND
Voorhees Township	0.2-0.4	ND	ND	0.1-0.2	ND	ND
Wenonah	1.2-1.6	ND	ND -0.1	0.1	ND	0.1
Westmont	0.1-0.2	ND	0.1-0.2	0.1-0.2	ND	ND

Westville	0.4–0.5	ND	ND –0.2	ND –0.1	ND	ND
Willingboro	1.0–1.3	ND	ND –0.1	0.1–0.2	ND	ND –0.1
Woodbury	1.2–2.1	ND	ND –0.1	0.1–0.4	ND	ND –0.2
Woodlynne	0.1	ND	ND –0.1	ND –0.2	ND	ND
Woodstown	1.9–2.5	ND	< 0.1	0.1–0.2	ND	ND

*denotes private water supply

ND denotes not detectable:

less than 0.1 µg F/ml
less than 0.01 µg Cr/ml
less than 0.01 µg Cu/ml
less than 0.01 µg Fe/ml
less than 0.03 µg Pb/ml
less than 0.01 µg Zn/ml

composite samples should be considered for analysis, or, preferably, a sampling frequency similar to that required for the evaluation of bacteriological quality (1) should be established.

Private domestic supplies are rarely evaluated for water quality on a routine basis. In the course of real estate transactions, certifications of water quality may be required, but these situations occur rather infrequently. In the case of private domestic supplies, public education programs stressing the need for periodic water quality evaluations coupled with laboratory services are needed.

Acknowledgements

The assistance of Norbert Ealer, Robert Jenkins and Craig Katz in making some of the laboratory measurements is gratefully acknowledged. Special thanks go to Patricia Colacci and Robert Smith for preparing the figures and to Rita Lorang for typing the manuscript.

Literature Cited

1. Potable Water Standards, New Jersey State Department of Environmental Protection, PW-D10, December 1970.
2. Standard Methods for the Examination of Water and Wastewater, 13th. ed., American Public Health Association, Washington, D.C., 1971.
3. Chloride Concentrations of Water from Wells in the Atlantic Coastal Plain of New Jersey, 1923-61, State of New Jersey, Department of Conservation and Economic Development, Special Report 22, 1963.
4. Okum, D.A., Water Management in England: A Regional Model, Environ. Sci. Technol., $\underline{9}$ 922 (1975).
5. Records of Wells and Ground Water Quality in Camden County, N.J., State of New Jersey, Department of Conservation and Economic Development, Water Resources Circular No. 10, 1963.

Separations of Biochemical Interest Using

Bonded Microparticle Ion Exchangers

Fredric M. Rabel

Whatman, Inc.

Clifton, N.J.   07014

The advances in high performance liquid chromatography (HPLC) have been arriving faster than most analytical chemists or biochemists can cope. These advances have been not only in the hardware, but also in the packings and columns. The result is that much confusion exists as to which system or components and which columns or packings would give the best results.

To reach any conclusions, the chromatographer has to know what is available commercially in terms of the hardware and packings, but also what his needs are now, and what they might be in the future. The purpose of this paper is to illucidate the advances in the column and packings technology and the uses of such packings in biochemical separations. This will at least give more background to the chromatographer who will be faced with making choices in the area of columns and packings in the future.

Liquid chromatography has become an essential element in the analysis of many biological constituents because of their polar nature and inherent solubility. As such, LC is growing in importance in clinical analysis. True, only clinical researchers are presently involved with HPLC, but within a few years liquid chromatography will surely find its way into the routine clinical laboratory. Details of the work presented here and discussion of possible advances will show the potential of this technique.

Because the chemistry of life relies on water soluble, ionizing compounds, the most used mode for biochemical sep-

arations has, of course, been ion exchange. Classically,
the large spherical, totally porous resinous ion exchangers
have provided the means to do many biological separations.
The disadvantage of this packing was the poor resolution
obtained and the time necessary for the separation. In the
mid-1960's, smaller spherical totally porous resinous ion
exchangers were available and first used quite successfully
for amino acid analysis. The amino acid analysers were,
in fact, the first modern liquid chromatographs. They did,
however, lack the speed necessary to move the LC technique
from limited to universal use.

    The need for faster mass transfer lead Horvath,
Lipsky [1-3], and Kirkland [4-8] to make thin layers of resinous ion exchangers bonded to glass beads, the first pellicular or surface bonded lc packings. The thin shell of
ion exchanger allowed fast equilibrium and exchange to move
lc into the era of high speed and high performance. The
first separation performed on the packings and published
were those of nucleic acid bases and nucleotides. Examples
of these separations are shown in Figures 1 and 2.

    Other researchers began working with these packings to
produce more sophisticated analyses of biological extracts.
Dr. Brown [9] and her co-workers have published widely in this
field, especially in the area of nucleotide analysis. The
pellicular packings offered the needed speed and efficiency
to allow many useful separations to be performed and certainly added impedious to the growth of liquid chromatography.

    In the 1960's most chromatographers were quite satis-
fied with the performance of the pellicular packings. The
theory, however, showed that even better columns should be
available if microparticle packings (20 micron or less) were
used. The secret of producing good microparticle columns,
however, was not solved until 1972. Researchers knew pressure packings of liquid (aqueous or nonaqueous) slurries
was necessary with the microparticles. Still particle segregation of such slurries during packing lead to non-homo-
geneously packed beds. The resulting columns gave poor ef-
ficencies and high back pressures. To alleviate the problems, the researcher needed a)narrow particle sized range
packings (10um $\pm$ 2um, not 5-20um material) and b)balanced
density solvents (to suspend the particles so they would not
segregate). Dr. Majors [10] was one of the first workers to
point out the importance of the combination of these two

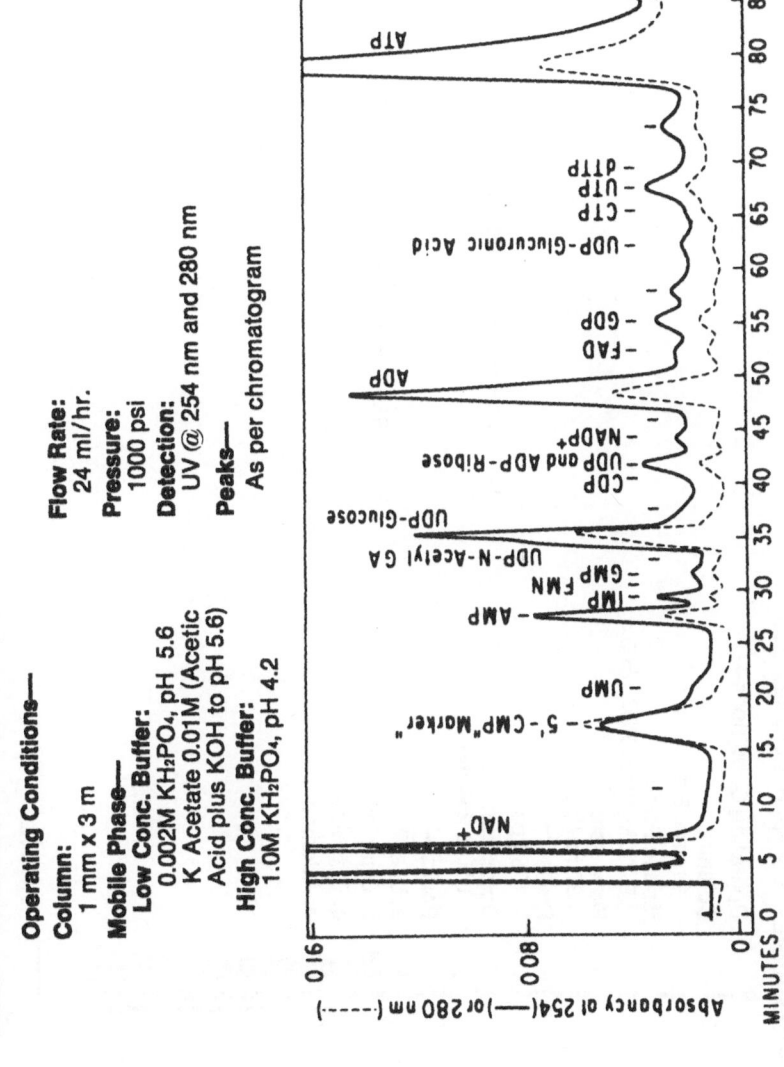

DIRECT LIVER EXTRACT OF 21 NUCLEOTIDES ON AS PELLIONEX SAX

Operating Conditions——

Column:
1 mm x 3 m

Mobile Phase——
Low Conc. Buffer:
0.002M KH₂PO₄, pH 5.6
K Acetate 0.01M (Acetic
Acid plus KOH to pH 5.6)

High Conc. Buffer:
1.0M KH₂PO₄, pH 4.2

Flow Rate:
24 ml/hr.

Pressure:
1000 psi

Detection:
UV @ 254 nm and 280 nm

Peaks——
As per chromatogram

Figure 1

BASES OF NUCLEIC ACIDS-PURINES AND
PYRIMIDINES ON HS PELLIONEX SCX

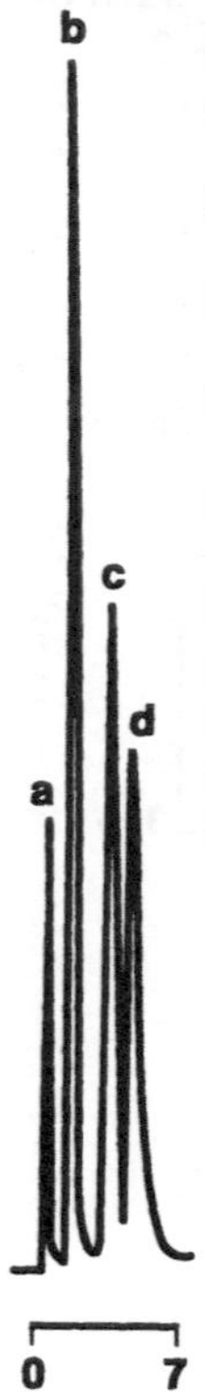

Operating Conditions——

Column:
1 mm x 3 m

Mobile Phase:
0.025M $NH_4H_2PO_4$, pH 3.5

Flow Rate:
37 ml/hr.

Pressure:
1750 psi

Detection:
UV @ 254 nm

Peaks——
a. Uracil
b. Guanine
c. Cytosine
d. Adenine

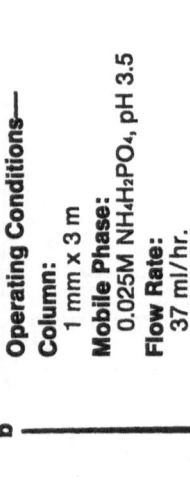

MINUTES

0   7

Figure 2

A. $Si-O-H + R-O-H \rightleftharpoons Si-O-R + H_2O$

B. $Si-O-H + Cl-Si-R_3 \longrightarrow Si-O-SiR_3 + HCl$

R = carbon chain

Figure 3   Bonding Chemistries for LC Packings

factors and to produce highly efficient microparticle columns.

At the same time as the packing technology was being developed, packing manufacturers were applying chemistry to bond various moieties to the silica microparticles. This bonding initially [11] was performed with alcohols which bonded to the Si-OH groups as shown in Fig. 3,A. The problem with the resulting bonds, however, was that they were not stable in water and alcohols - the bonding being reversible in these solvents. Researchers [12-15] next produced bonds with organosilanes so that Si-O-Si-C bonds were formed (Figure 3,B). Such products were thermally and hydrolytically stable and are the basis for all packings to be discussed. The phases which could be bonded are infinite, but because of the importance of the mobile phase as a variable in optimizing a separation, only a few bonded phase microparticle packs appear to be necessary for most separations.

Obviously the chromatographer needed ion exchangers bonded to microparticles to do the most efficient analysis of biological extracts. The Partisil bonded microparticle exchangers were manufactured by Whatman in 1974 to fill this need. The basic structure of these is shown in Table I with their ionogenic groups. It should be noted that different

Table I

Microparticle Ion Exchange Packings

Product	Ionogenic Group	Counter Ion
Partisil-10 SAX	———— * $\overset{+}{N}R_3$	$H_2PO_4^-$
Partisil-10 SCX	———— * ⬡ SO_3^-	NH_4^+

* proprietary

ion exchange properties will be found in the various ion
exchange packings manufactured from company to company.
This is because of the chemistries in the bonding process
vary to a large extent. Thus even though the ionogenic
group may be the same, the chain length holding the group to
the support, amount of cross linking, or amount of bonded
phase will vary with each company to give different chromatographic properties to the resulting product. For instance,
the AS Pellionex SAX has a styrene-divinyl benzene backbone
bonded to the glass bead. The Partisil-10 SAX has an aliphatic chain bonded to the silica. Both are terminated by
the same quaternary nitrogen group but the selectivities are
different for the monophosphate nucleotides as shown in Figure 4 (as compared to Figure 1). Figure 5 shows the deoxynucleotides also on Partisil-10 SAX, and Figure 6, the
bases on Partisil-10 SCX.

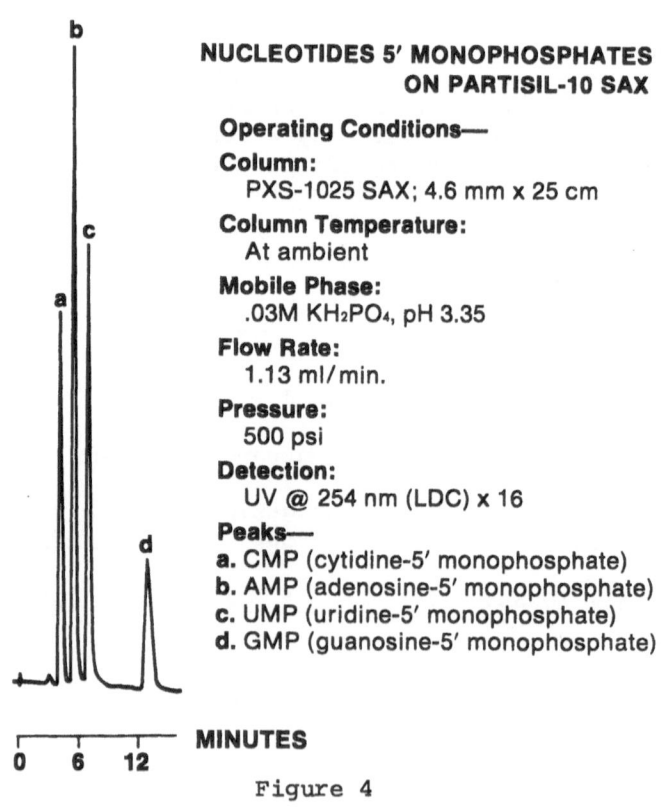

NUCLEOTIDES 5' MONOPHOSPHATES ON PARTISIL-10 SAX

Operating Conditions—

Column:
PXS-1025 SAX; 4.6 mm x 25 cm

Column Temperature:
At ambient

Mobile Phase:
.03M KH_2PO_4, pH 3.35

Flow Rate:
1.13 ml/min.

Pressure:
500 psi

Detection:
UV @ 254 nm (LDC) x 16

Peaks—
a. CMP (cytidine-5' monophosphate)
b. AMP (adenosine-5' monophosphate)
c. UMP (uridine-5' monophosphate)
d. GMP (guanosine-5' monophosphate)

MINUTES

0   6   12

Figure 4

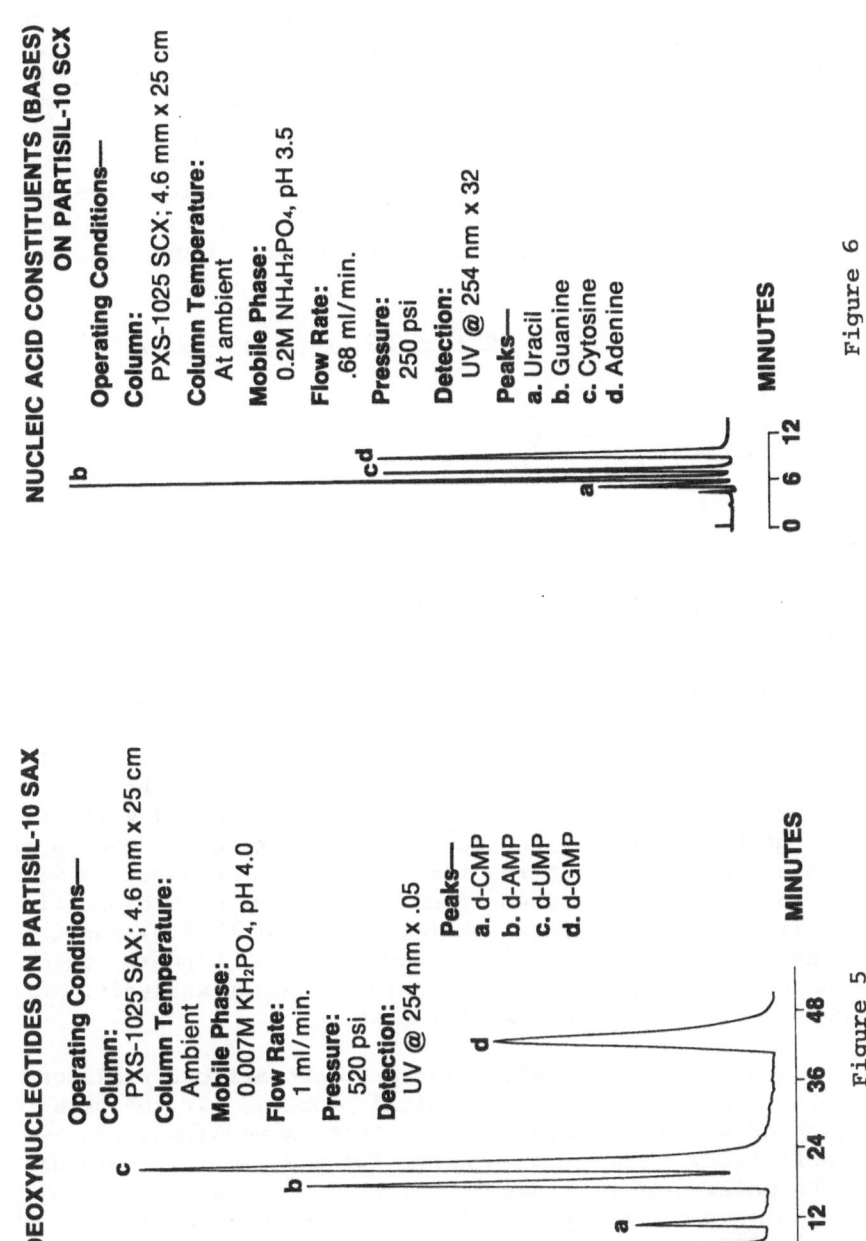

NUCLEIC ACID CONSTITUENTS (BASES) ON PARTISIL-10 SCX

Operating Conditions—

Column:
PXS-1025 SCX; 4.6 mm x 25 cm

Column Temperature:
At ambient

Mobile Phase:
0.2M NH4H2PO4, pH 3.5

Flow Rate:
.68 ml/min.

Pressure:
250 psi

Detection:
UV @ 254 nm x 32

Peaks—
a. Uracil
b. Guanine
c. Cytosine
d. Adenine

MINUTES

Figure 6

DEOXYNUCLEOTIDES ON PARTISIL-10 SAX

Operating Conditions—

Column:
PXS-1025 SAX; 4.6 mm x 25 cm

Column Temperature:
Ambient

Mobile Phase:
0.007M KH2PO4, pH 4.0

Flow Rate:
1 ml/min.

Pressure:
520 psi

Detection:
UV @ 254 nm x .05

Peaks—
a. d-CMP
b. d-AMP
c. d-UMP
d. d-GMP

MINUTES

Figure 5

In order to choose which ion exchange media to use, the researcher has to look at the properties of the solutes to be separated. If the solutes give acidic solutions, indicating release of protons, and a resulting anionic species, then an anion exchanger is required for the separation of components with such chemical properties. Conversely, if the solutes give basic solutions, indicating release of hydroxyl groups and a resulting cationic species, then a cation exchanger is required for the separation (See Figure 7).

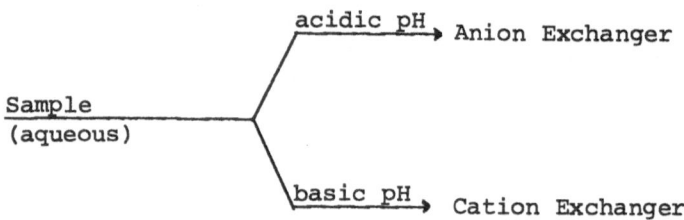

Figure 7   Selection of an Ion Exchange Column

Some biochemicals are amphoteric, therefore could be separated on either cation or anion exchangers. Usually one exchanger will work better for the separation because of better resolution or the ability of one exchanger to better work around any co-extractants. Only experimental work can show the best possible choice. In addition to the choice of anion or cation exchangers, the researcher can elect between the use of pellicular (superficially porous) or microparticle (totally porous) packings. Eliminated from this discussion is the microparticle totally porous resin exchangers, which although widely used for amino acid and other analysis, still lack the high efficiencies of the above.

Choosing between pellicular or microparticle packings involves considering the analytical problem. If the separation of a few components is all that is required, the lower efficiency ($\sim$ 500 plates/meter for a 2.1mm x 1m column) of the pellicular packings is all that are required. The pellicular packings are easily dry-packed with a vibrator [16] or with tap-packing [17]. Pellicular columns and pellicular pre-columns are recommended if dirty samples or extracts are to be placed on the LC column since replacement is easy.

Pellicular pre-columns should be used before both pellicular
and microparticle analytical columns because they are most
readily made.   The use of pre-columns will insure added life
of the analytical column [18].

If a more difficult or multi-component sample needs to
be analyzed, the more efficient ($\sim$ 10,000 plates/meter for a
4.6mm x 25cm column) microparticle Partisil column would be
required.   Because of the higher efficiency of the Partisil
ion exchange columns, more care in use is recommended to
maintain maximum performance.   Many suggestions have been
given in a recent article by this author [18] and will not be
elaborated upon at this time.

To insure the high efficiency on the Partisil columns,
only a monomolecular layer of ion exchanger is built onto
the microparticle support.   Because of the nature of the
microparticle support of most presently available packings -
namely, silica, such monomolecular layers do not protect the
support from chemical action by basic (pH > 7.5) solutions.
Thus all silica microparticle packings (bonded and non-
bonded) unless recommended otherwise by the manufacturers
should not be used with high pH buffers.   Most pellicular
packings (on glass beads) however, are completely resistant
to high pH, and are the best choice for separations requiring such pH's.   The future will no   doubt bring more resistant microparticle packings.

Getting the Initial Chromatogram

After a logical choice of column has been made, the
column is packed or purchased and attached to the liquid
chromatograph.   The system has to be purged until equilibrium
is established with a suitable buffer.   Initially, it is best
to elute all of the sample through the column.   This means use
a fairly strong buffer if running isocratically (single
solvent or buffer) or a gradient to a strong buffer for the
initial chromatogram.

Each manufacturer will have salts and concentrations
best used with his products.   Whatman pellicular (Pellionex)
and microparticle (Partisil) packings and columns work well
with phosphate buffers (sodium, potassium, ammonium; mono-
basic, dibasic, tribasic phosphate) in the concentration
range of 0.005 to 0.5M for most separations.   Other buffers'
can be used to effect different selectivities.

The initial pH of the buffer should be known and can
be any pH so that the solutes of interest are in an ionic
form and will so remain when on the column. pH optimization
follows the adjustment of the ionic strength discussed below.

First adjust the ionic strength (either stepwise or
with gradient elution) to that which will elute the solutes
with in a k' range of from 2-10. If doing stepwise elution,
the buffer strength is slowly decreased, allowing about 30ml
of each new buffer to re-equilibrate the column before re-
injecting the sample.

After a few peaks are found in the proper k' region,
the pH of the buffer is changed to see if improvement in the
resolution is possible. The pH is changed 0.5 units in each
direction from the initial pH. When equilibrating the column
to a new pH, more equilibrating solvent ($\sim$200ml) may be required since pH equilibrium takes longer to establish than
does ionic strength equilibrium. Changing the pH is a powerful tool in biochemical separations. Peaks can be readily
shifted about the chromatogram by controlling the amount of
solute ionization. If a solute is un-ionized at a given pH,
it will most likely elute at the t_0 of the column. pH adjustment is a means of eliminating interferring compounds,
if the pka's of the solutes and impurities are different
enough from each other.

Because partition chromatography occurs in all ion exchange separations (the organic structure supporting the
ion exchange sites is an ideal partitioning media) it can
also be made to play a more important role in biochemical
separations. To optimize on the partition chromatography,
some alcohol (methyl, ethyl, propyl) should be added -
stepwise to a maximum of 10% - to the mobile phase. This
addition of alcohol can help, hinder, or do no harm to the
separation, thus has to be tried for each new separation.

The sum total of varying the above parameters will be
a very selective separation, many times possible only with
ion exchange chromatography. Some of the possible separations accomplished on the Partisil SAX and SCX columns are
shown in Figures 8-12. In all cases, the procedure as outlined above was used to develop the finished chromatogram.

Figure 8 and 9 are separations of antibiotics on Par-

tisil-10 SAX. These, of course, are probably more important
as presented to those involved in the manufacture of pharmaceuticals. Perhaps, however, LC will offer a convenient
method for following drug treatment and dosage. As is usual,
the problems involved with the use of physiological fluids
will be the extraction and concentration techniques. Good
experimental work will be necessary to refine these techniques. The separation itself will not present problems
for the most part because of the increased efficiency to the
microparticle packings.

Figure 10 on catecholamines is of greater interest
since these compounds are being extensively researched because of their importance as neurotransmitters. Other compounds of this group of compounds are also readily separated
on this Partisil-10 SCX column.

Figure 11 is of two UV absorbing amino acids. Because
of the higher efficiency of the Partisil-10 SCX a complete
amino acid profile should be possible in a much reduced
time scale as has been possible with the conventional microparticle resinous cation exchangers.

Figure 12 shows the separation of vitamin B_6 into three
of its components on Partisil-10 SCX. This same packing has
been used to separate a mixture of sodium ascorbate, pyridoxine hydrochloride, niacinamide, riboflavin and thiamine
mononitrate with a mobile phase of $0.1\underline{M}$ $(NH_4)_2HPO_4$, pH 2.2 [19].

Alternatives to Ion Exchange Chromatography

Although ion exchange is a powerful separation tool for
biochemical samples, success is also possible through the use
of other chromatographic modes. In particular, the partition
mode using bonded non-polar (C18) and polar (cyano) groups on
the microparticles has also proven to be useful. Whether the
column used in the Partisil-10 ODS (C18) or the Partisil-10
PAC (cyano) depends upon the polarity of the solutes in the
mixture. The old chemical adage 'like dissolves like' is
important here.

In addition to the need for the correct polarity bonded
phase, the use of these modes is only possible by performing
the separation on associated molecules. If the biological
species are allowed to ionize, the result will be a chromato-

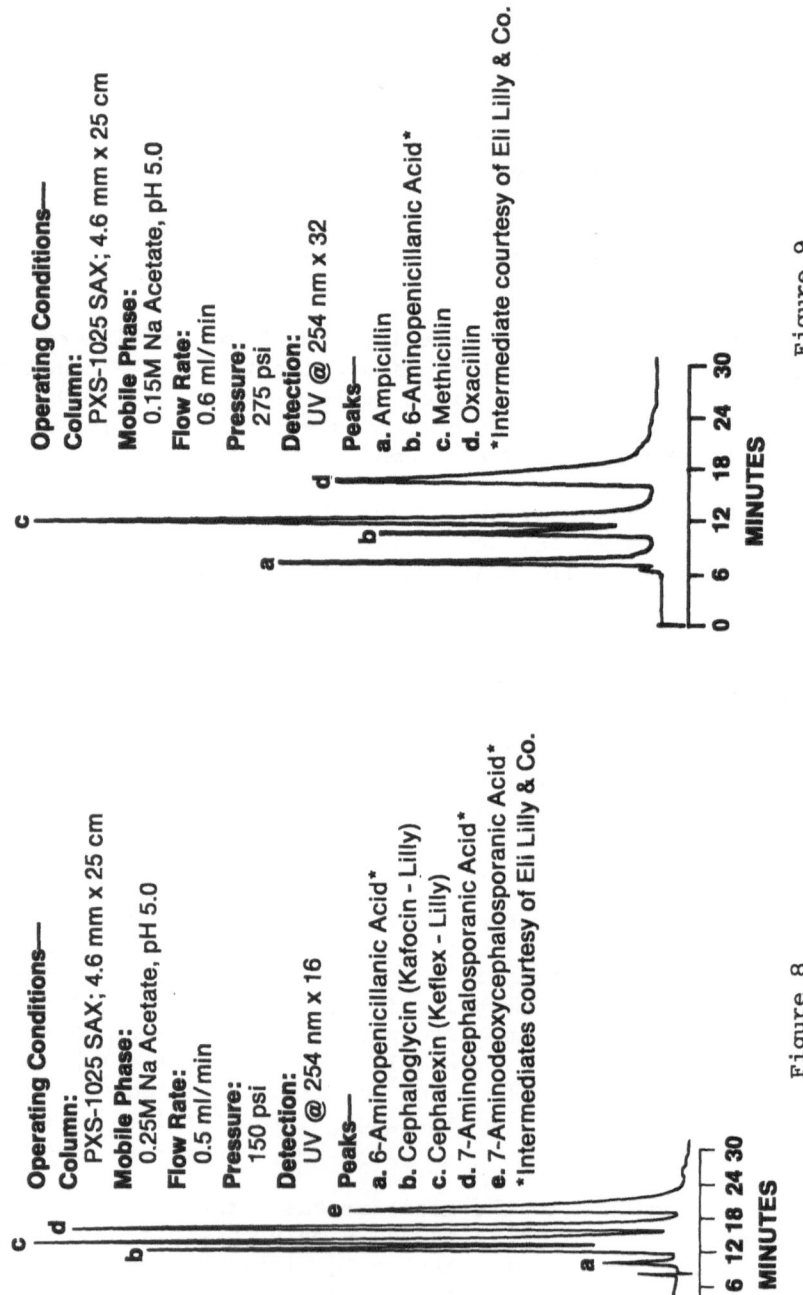

SYNTHETIC PENICILLINS ON PARTISIL-10 SAX

Operating Conditions—
Column:
PXS-1025 SAX; 4.6 mm x 25 cm
Mobile Phase:
0.15M Na Acetate, pH 5.0
Flow Rate:
0.6 ml/min
Pressure:
275 psi
Detection:
UV @ 254 nm x 32
Peaks—
a. Ampicillin
b. 6-Aminopenicillanic Acid*
c. Methicillin
d. Oxacillin
*Intermediate courtesy of Eli Lilly & Co.

MINUTES

Figure 9

CEPHALOSPORINS ON PARTISIL-10 SAX

Operating Conditions—
Column:
PXS-1025 SAX; 4.6 mm x 25 cm
Mobile Phase:
0.25M Na Acetate, pH 5.0
Flow Rate:
0.5 ml/min
Pressure:
150 psi
Detection:
UV @ 254 nm x 16
Peaks—
a. 6-Aminopenicillanic Acid*
b. Cephaloglycin (Kafocin - Lilly)
c. Cephalexin (Keflex - Lilly)
d. 7-Aminocephalosporanic Acid*
e. 7-Aminodeoxycephalosporanic Acid*
*Intermediates courtesy of Eli Lilly & Co.

MINUTES

Figure 8

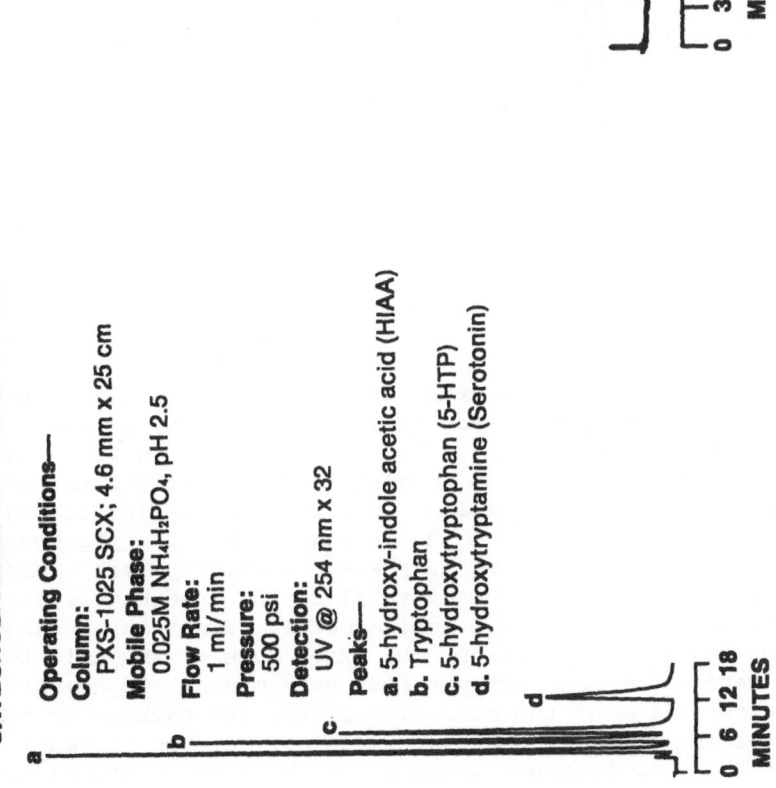

AMINO ACIDS ON PARTISIL-10 SCX

Operating Conditions—
Column:
PXS-1025 SCX; 4.6 mm x 25 cm
Mobile Phase:
.05M KH$_2$PO$_4$, pH 4.50
Flow Rate:
0.5 ml/min
Pressure:
380 psi
Detection:
UV @ 254 nm x .10
Peaks—
UV absorbing amino acids
a. Tyrosine
b. Phenylalanine

Figure 11

CATECHOLAMINES ON PARTISIL-10 SCX

Operating Conditions—
Column:
PXS-1025 SCX; 4.6 mm x 25 cm
Mobile Phase:
0.025M NH$_4$H$_2$PO$_4$, pH 2.5
Flow Rate:
1 ml/min
Pressure:
500 psi
Detection:
UV @ 254 nm x 32
Peaks—
a. 5-hydroxy-indole acetic acid (HIAA)
b. Tryptophan
c. 5-hydroxytryptophan (5-HTP)
d. 5-hydroxytryptamine (Serotonin)

Figure 10

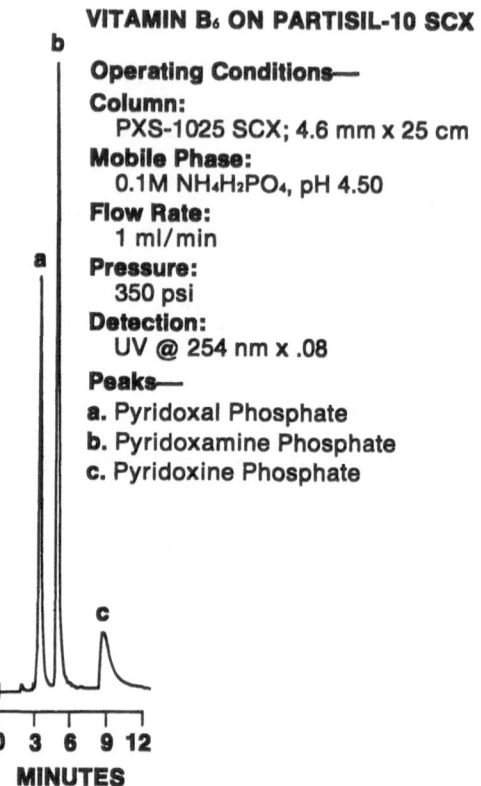

VITAMIN B₆ ON PARTISIL-10 SCX

Operating Conditions—
Column:
  PXS-1025 SCX; 4.6 mm x 25 cm
Mobile Phase:
  0.1M $NH_4H_2PO_4$, pH 4.50
Flow Rate:
  1 ml/min
Pressure:
  350 psi
Detection:
  UV @ 254 nm x .08
Peaks—
a. Pyridoxal Phosphate
b. Pyridoxamine Phosphate
c. Pyridoxine Phosphate

Figure 12

gram with badly tailing peaks and poor resolution. To force
the ionized species to associate, acids, bases, or salts are
added to the mobile phase. This is analogour to adding acid
or base to the developing solvent in TLC to eliminate tailing
of spots. Thus the excellent separation of analgesics on
Partisil-10 ODS in Figure 13 is possible because phosphoric
acid was added to the mobile phase. Similarly, the carbohydrates separated on the Partisil-10 PAC (see Figure 14)
required this acid in the mobile phase. Since carbohydrates
are un-ionized in solution, the exact role of the acid in
this separation has not yet been determined. Apparently,
ion association of the acid with either the packing or solutes is important. Some species do not require reagents
other than the methanol/water as can be seen in the barbitu-
rate separation in Figure 15.

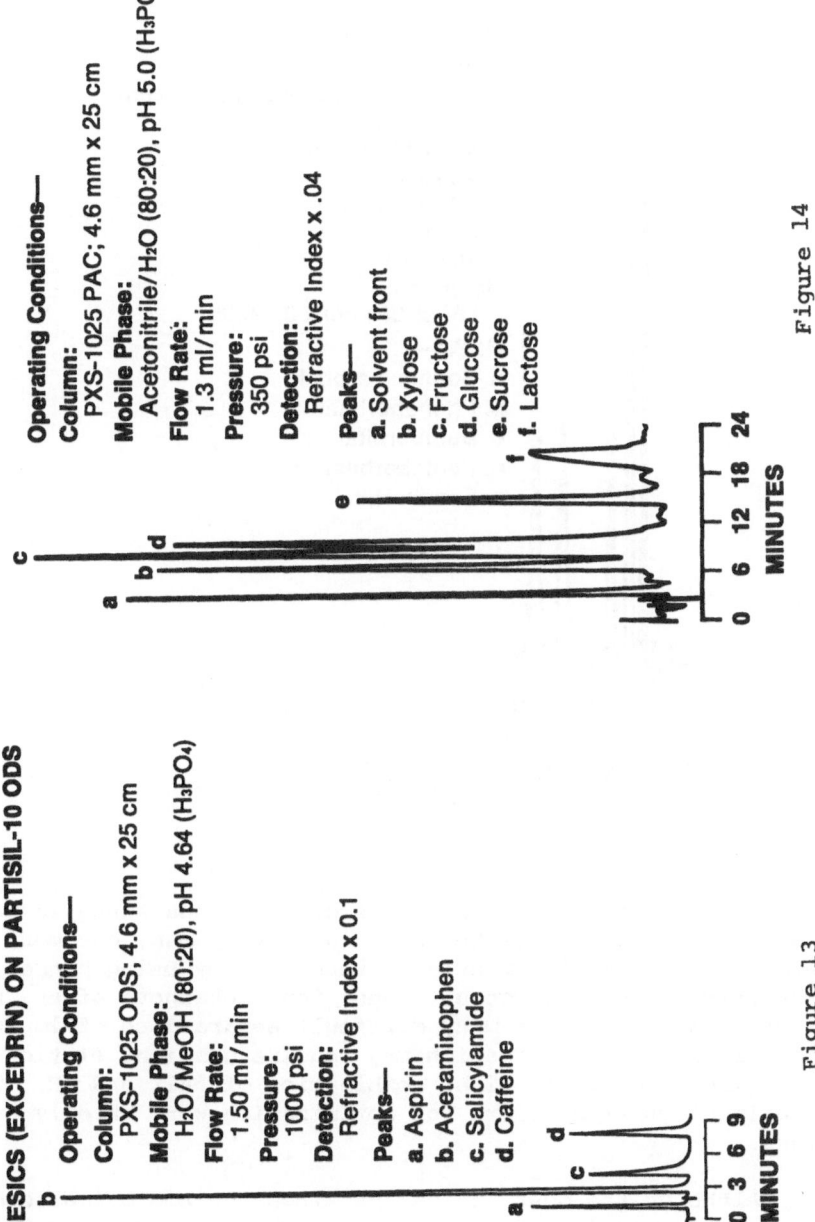

ANALGESICS (EXCEDRIN) ON PARTISIL-10 ODS

Operating Conditions—
Column:
  PXS-1025 ODS; 4.6 mm x 25 cm
Mobile Phase:
  H₂O/MeOH (80:20), pH 4.64 (H₃PO₄)
Flow Rate:
  1.50 ml/min
Pressure:
  1000 psi
Detection:
  Refractive Index x 0.1
Peaks—
  a. Aspirin
  b. Acetaminophen
  c. Salicylamide
  d. Caffeine

Figure 13

CARBOHYDRATES ON PARTISIL-10 PAC

Operating Conditions—
Column:
  PXS-1025 PAC; 4.6 mm x 25 cm
Mobile Phase:
  Acetonitrile/H₂O (80:20), pH 5.0 (H₃PO₄)
Flow Rate:
  1.3 ml/min
Pressure:
  350 psi
Detection:
  Refractive Index x .04
Peaks—
  a. Solvent front
  b. Xylose
  c. Fructose
  d. Glucose
  e. Sucrose
  f. Lactose

Figure 14

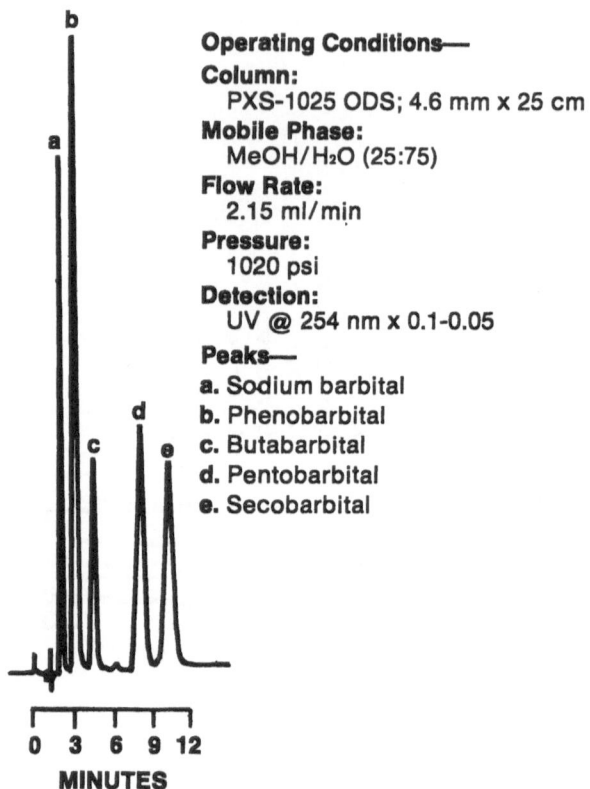

BARBITURATES ON PARTISIL-10 ODS

Operating Conditions—
Column:
  PXS-1025 ODS; 4.6 mm x 25 cm
Mobile Phase:
  MeOH/H₂O (25:75)
Flow Rate:
  2.15 ml/min
Pressure:
  1020 psi
Detection:
  UV @ 254 nm x 0.1-0.05
Peaks—
  a. Sodium barbital
  b. Phenobarbital
  c. Butabarbital
  d. Pentobarbital
  e. Secobarbital

Figure 15

Summary

     The separation of many types of compounds found in the
human body, produced either by nature or by man, can more
readily and more efficiently be separated by using bonded microparticle ion exchangers. These ion exchangers offer the
needed selectivity for many difficult separations of ionized
biologicals. The separation may require gradient elution or
many times simply the right combination of salt and pH. The
details of this procedure for optimization were briefly outlined.

     Although the most logical packings for biological com-
pounds which ionize would seem to be the ion exchangers, the
use of other bonded phases has been briefly discussed. These

microparticle bonded phases combined with good extraction and clean up techniques should result in greater sensitivity of detection of pertinent molecules in many metabolic pathways. Hopefully this in turn will lead to better treatment or preventative measures relating to diseased states.

References

1) Horvath, C. and Lipsky, S.R., Anal.Chem.,39,1442 (1967)
2) Horvath, C. and Lipsky, S.R., Anal.Chem.,41,1227 (1969)
3) Horvath, C. and Lipsky, S.R., J.Chrom.Sci., 7,109 (1969)
4) Kirkland, J.J., J. Chrom.Sci., 7,7 (1969)
5) Kirkland, J.J., J. Chrom.Sci., 7,361 (1969)
6) Kirkland, J.J., J. Chrom.Sci., 8,73 (1970)
7) Kirkland, J.J., J.Chrom.Sci., 10,129 (1972)
8) Kirkland, J.J. & DeStefano,J.J.,J.Chrom.Sci.,8,309 (1970)
9) Brown,P.R., High Pressure Liquid Chromatography, Academic Press, New York, 1973.
10) Majors, R.E., Anal.Chem 44, 1722 (1972)
11) Halasz, I. and Sebastian, I., Angew.Chem., Internat. Ed., 8, 453 (1969)
12) Aue, W.A., and Hastings, C.R., J.Chrom.,42,319 (1969)
13) Brust, E.E., Sebastian, I., and Halasz, I., J.Chrom., 83,15 (1973)
14) Novontny, M., Bektesh, S.L., and Grohmann, K., J.Chrom., 83, 25 (1973)
15) Locke, D.C., Schmurmund, J.T., and Banner, B., Anal.Chem., 44, 90 (1972)
16) Chroma-Tap, available from Whatman, Inc.
17) Snyder, L.R. and Kirkland, J.J., Introduction to Modern Liquid Chromatography, Wiley, New York, 1974, p.189.
18) Rabel, F.R., Amer.Lab., 7, 53 (1975), May issue (No.5).
19) Private Communication from Pennex Products Co., Inc.

PROBLEMS OF LIQUID SAMPLING IN

GAS CHROMATOGRAPHY

L.S. Ettre and J.E. Purcell

The Perkin-Elmer Corporation

Norwalk, Connecticut 06856

INTRODUCTION

A gas chromatograph is a very complex system. Of course, its heart is the column in which the separation takes place; however, the true utilization of the column's separation ability depends on the whole system, on the optimum operation of each components.

In spite of this, some parts of the system are rarely discussed and their proper functioning is more-or-less assumed. Also, while we take it for granted that, depending on the separation problem, columns with various stationary phases have to be selected, we usually forget that other parts of the system also have to be selected according to the analytical problem. For example, the optimization of high efficiency columns, or the detection of sub-picogram quantities is discussed exhaustively in the literature; however, the construction, selection and optimization of the proper sample inlet systems are the subject of a very limited number of papers.

The purpose of this paper is to fill this gap, by discussing the problems associated with liquid sampling. First, we shall survey the various injection systems demonstrating that there is no universal system. Subsequently, the problems associated with the common septa used in these sampling systems will be discussed and possibilities for their elimination described. Both discussions will rely on work

carried out over the past twenty years using systems developed by Perkin-Elmer to illustrate the various points.
Questions associated with liquid sampling were summarized in 1968 by Condon and Ettre (1) and the present text
may be considered as a continuation of that book chapter,
while the discussion of the problems associated with the
common septa represents a summary of a more detailed
paper published recently (2).

The discussion will be restricted to manual systems; in
other words, the special criteria and problems of semi-
automated and automated systems are not the subject of
this paper. Also, only systems using a septum and a syringe for sample injection will be mentioned. Of course,
other systems also exist; however, there is no question
that the majority of present-day gas chromatographs still
utilize septum and syringe.

As mentioned, sample pick-up and introduction is done
with the help of syringes. Early syringes used at the beginning of the development of GC had many drawbacks related mainly to their fairly large volume. The real break-
through occurred around 1958 when C.H.Hamilton designed
and began to manufacture his precision microsyringes which
really revolutionized sample handling. Without these devices we would not have the advanced sampling systems
common today.

SYSTEMS FOR LIQUID SAMPLE INTRODUCTION

In their first paper on gas-liquid partition chromatography
published in 1952 (3) James and Martin used a micropipet
for sample introduction. The carrier gas flow was interrupted, the column disconnected, the sample applied to
the top of the column and then, the flow was resumed. Two
years later, in 1954, Ray simplified the method of liquid
sample introduction (4). He applied a serum bottle cap on
the top of the column and introduced the liquid sample with
the help of an Agla micrometer syringe. From then on, injection of the liquid sample through a rubber-type seal
with help of a syringe became the universally applied technique of liquid sample introduction. Within one year of

Ray's work, the system was further improved resulting in the heated flash vaporizers of the first commercial gas chromatographs (5).

Flash Vaporizers

As shown in Fig.1/a, these early systems consisted of a metal block and a rubber septum which was compressed in a metal fitting. The block was either heated by an external heating plate or a cartridge heater was imbedded in the metal wall. At that time all gas chromatographs permitted only isothermal operation and thus, the carrier gas was usually preheated by being conducted to the injection block through the oven of the instrument. Since the temperature of the injection block was generally higher than that of the oven, the temperature of the carrier gas was always less than the injector temperature. Principally, this may have presented a problem: however, since the early instruments were rarely used at temperatures above 180°C, this problem was not evident.

The development of the GC analysis of fatty acid methyl esters on polyester columns suddenly adapted the technique to a very important application where column temperatures of 200-220°C were used and the sudden evaporation of the sample required even higher injection block temperatures. For this reason, the construction of injection blocks was changed so that the cartridge heater was placed inside the evaporation zone. This construction is shown in Fig. 1/b. However, later experience showed that while this modified construction provided the necessary heat capacity for instantaneous evaporation, sample decomposition could readily occur due to a non-uniform temperature within the evaporation zone.

A further problem of early injection system designs can be demonstrated by the all-glass systems introduced in the early nineteen sixties. In the early systems shown in Fig. 2/a, the top of the glass column was essentially construck-ted as a tee which was surrounded by a heated metal block, and a cylinder-sized silicone rubber plug was used in lieu of the septum. The problem soon evident with these systems

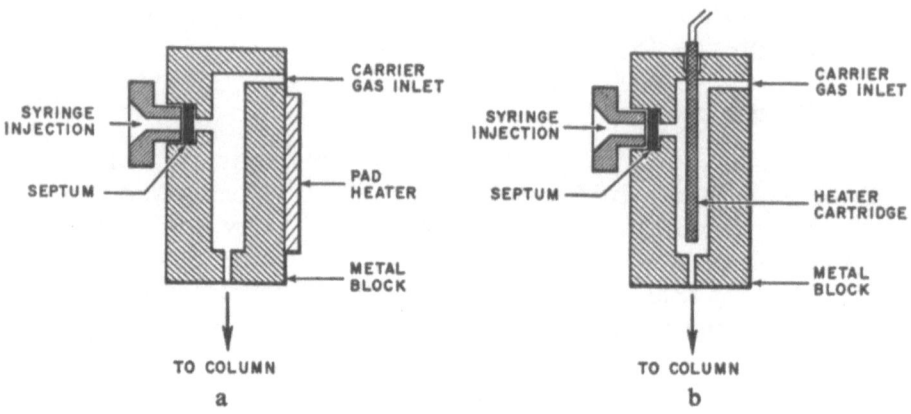

Fig.1. Simplified schematics of various types of injection blocks ("flash vaporizers") of early design.
(a) Block heated externally; (b) cartridge heater inside the evaporation zone.

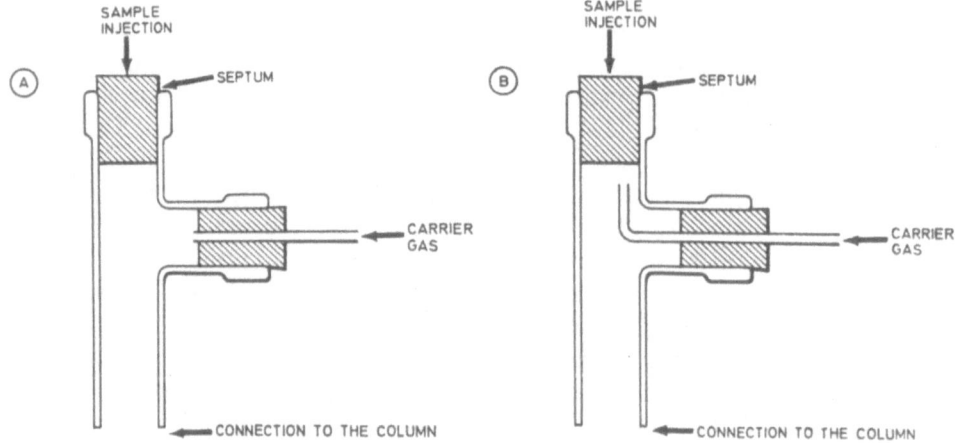

Fig.2. Two types of early systems for direct injection into glass packed columns. The top of the column is surrounded by a heated metal block (not shown).
(a) Basic design; (b) septum area flushed with the carrier gas.

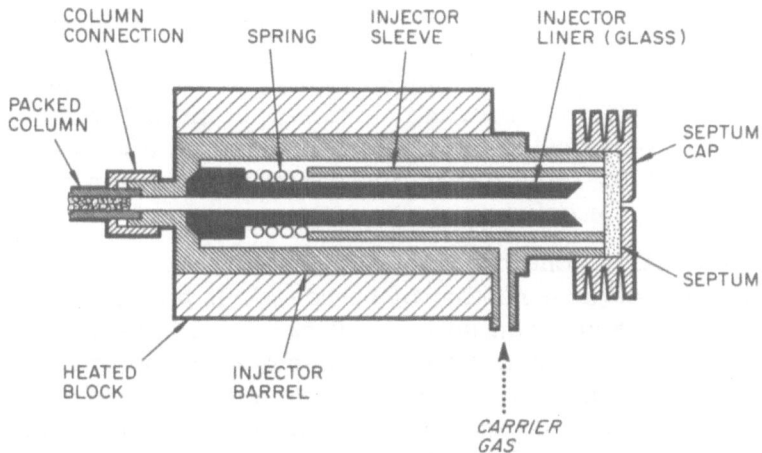

Fig.3. Simplified schematic of a sample introduction system
with improved carrier gas flow pattern.

was a fairly large unswept volume close to the septum
wherein part of the evaporated sample accumulated and
later diffused back into the carrier gas stream. A modification shown in Fig.2/b represents a partial remedy to this
problem.

We have described three problems associated with inadequate sample introduction system design: lack of carrier
gas preheating, non-uniformity of the temperature in the
evaporation zone, and dead-volumes within the system.
Improvements in the flash-vaporizer type sample introduction systems were aimed at eliminating these problems. A
number of systems were developed in the nineteen sixties
each representing an advancement as compared to the previous design (1). Fig.3 shows the simplified schematic of
the most advanced, up-to-date system. Let us discuss it
in detail.

In this design, the carrier gas is preheated to the injection
block temperature, by having a sufficiently long path inside the block itself before coming into contact with the

sample vapor. The septum is kept somewhat cooler than
the block itself and the volume immediately adjacent to it
is efficiently swept with high-velocity carrier gas. The
whole flow pattern within the system is designed to minimize back-diffusion to the cooler septum area and provi-
sion is made for rapid expansion and transfer of the sample
vapor toward the column.

The evaporation zone proper is provided by a glass insert.
This insert serves a number of purposes. It provides a relatively inert and clean surface. It has been shown by a
number of workers investigating sample decomposition that
the most critical moment is during evaporation, and that
decomposition can be induced not only by metal but also by
charred residues of past samples or by hot spots within the
evaporation zone. These inserts can be easily removed and
cleaned with a simple pipe cleaner or by dipping them in a
solvent. Also, the thermal nature of glass provides a more
uniform heat distribution as compared to metal.

Utilization of such inserts also permits us to adjust the
volume of the evaporation chamber, by the proper selection
of the inside diameter of the glass insert, without the need
to replace the entire injection system. Let us stop at this
point for a moment.

In case of direct sample introduction via evaporation, our
aim is to transfer the sample vapor into the column as a
slug with a minimum of diffusion. Here, we can have two
problems: the volume of the evaporation chamber can be
too large or too small.

The first case is represented by an evaporation chamber
having a volume much greater than the vapor volume of the
sample. In this case, band-spreading due to diffusion
within the chamber will occur reducing the separation between peaks. This effect can be demonstrated by direct
sample injection into support-coated open tubular (SCOT)
columns, utilizing evaporation chambers of different volumes. Two such chromatograms are shown in Figs. 4/b-c
(6). The sample volume was identical (0.2 μl) in both cases
while the internal volume of the glass liner in case of

Table I. Components Present in the Sample Analyzed For
Figures 4/a-c.

Peak No.	Substance	Boiling Point, °C
1	Methyl chloride	− 24.2
2	Vinyl chloride *	13.4
3	Ethyl chloride *	12.3
4	Dichloromethane	40.0
5	1,2-Dichloroethylene, trans	47.5
6	?	
7	1,2-Dichloroethylene, cis	60.3
8	Trichloromethane	61.7
9	Tetrachloromethane	76.5
10	1,2-Dichloroethane	83.5
11	1,2,2-Trichloroethylene	87.0
12	?	
13	1,1,2,2-Tetrachloroethylene *	121.0
14	1,1,2-Trichloroethane *	113.8

* There is some uncertainty concerning the sequence of
the compound pairs corresponding to peaks 2-3 and 13-14.

Fig.4/c was four times that of Fig.4/b. The sample consisted of technical 1,2-dichloroethane (95+%) with over a
dozen organochlorine impurities (cf. Table I).

The detrimental effect of the excessively large evaporation
chamber in case of Fig.4/c is evident from the drastic reduction of the number of theoretical plates calculated for
peak No.9 and the loss of resolution of peaks 4,5 and 6
and of peaks 10 and 11.*

Naturally, direct injection into a SCOT column represents
a compromise and the preferred mode is split-flow injec-

---

* It should be noted that the resolution of peaks 12, 13 and
14 remained almost identical. The reason for this is that
the temperature programming in the second half of the
chromatogram reduced the effect of band-spreading for
these later peaks. This is a well known phenomenon.

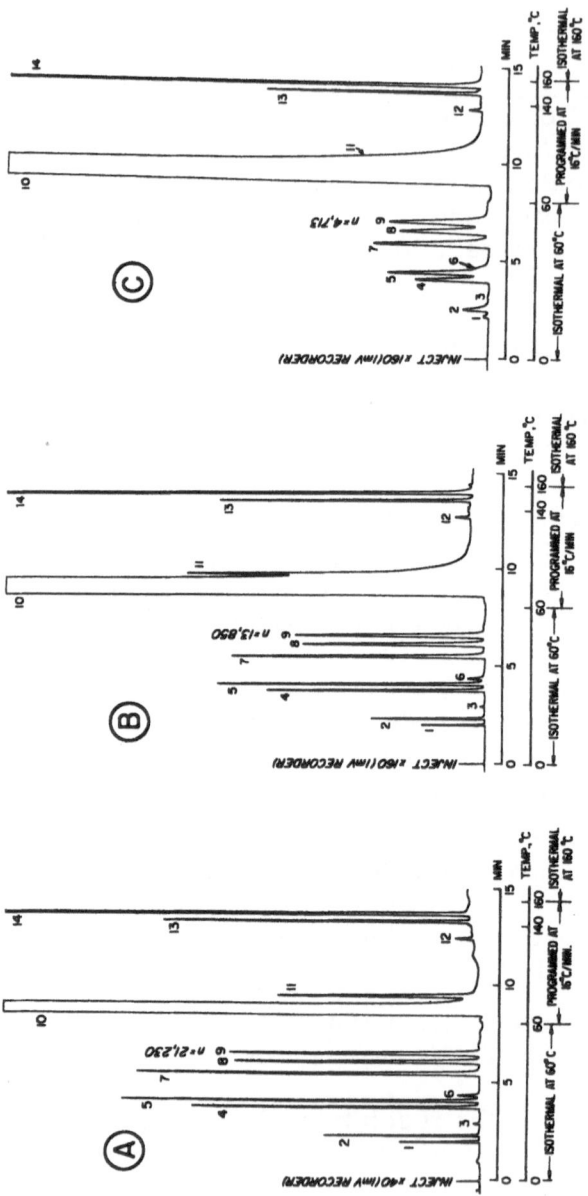

Fig. 4. Analysis of a chlorinated hydrocarbon mixture on a 50 ft x 0.020 in. ID SCOT column. Liquid phase: m(bis–m–phenoxyphenoxy)benzene + Apiezon L. Injector temperature: 250°C. FID temperature: 250°C. Carrier gas (He) flow rate: 4 ± 0.4 ml/min. For peak identification see Table I.

(a) Split injection, 1/33 split ratio. Sample volume injected 3.2 µl, entering the column 0.1 µl.

(b) Direct injection (no split); 1.5 mm ID glass liner. Sample volume 0.2 µl.

(c) Direct injection (no split); 3.0 mm ID glass liner. Sample volume 0.2 µl.

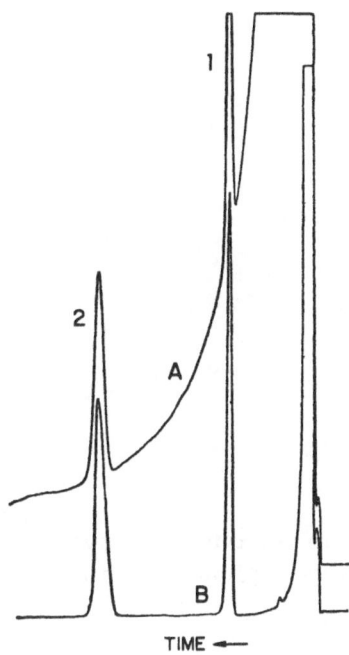

Fig.5. The effect of the volume of the evaporation chamber on the analysis.
Column: 2% H_3PO_4 + 20% diethyleneglycol adipate on Chromosorb W 80/100 mesh.. Column temperature: 120°C. Carrier gas (He) flow rate: 30 ml/min. FID. Peaks: 1 methyl octanoate, 2 methyl decanoate. Sample volume: 3.4 µl.
(a) 1.5 mm ID glass liner;
(b) 3.0 mm ID glass liner.

tion. This is evident if we compare Figs.4/b-c with Fig. 4/a obtained by using the optimum sampling for such columns: again, compare the number of theoretical plates and the resolution of the above listed peaks.

The second problem is represented by an evaporation chamber having an insufficient volume as compared to the vapor volume of the sample. This is illustrated in Fig.5. The sample consisted of an n-hexane solution of about 100 ppm each of methyl octanoate and decanoate, and the same volume was injected in both cases into glass liners having different internal diameters. As seen, in the first case, with the narrow bore liner, the solvent peak was so broad that it overlapped both component peaks while in the second case (liner with larger internal diameter), a sharp and well-separated solvent peak was obtained (6).

As a conclusion of these chromatograms it can be stated that there is no "universal" sample inlet system, not even in the case of the relatively simple flash vaporizers; the

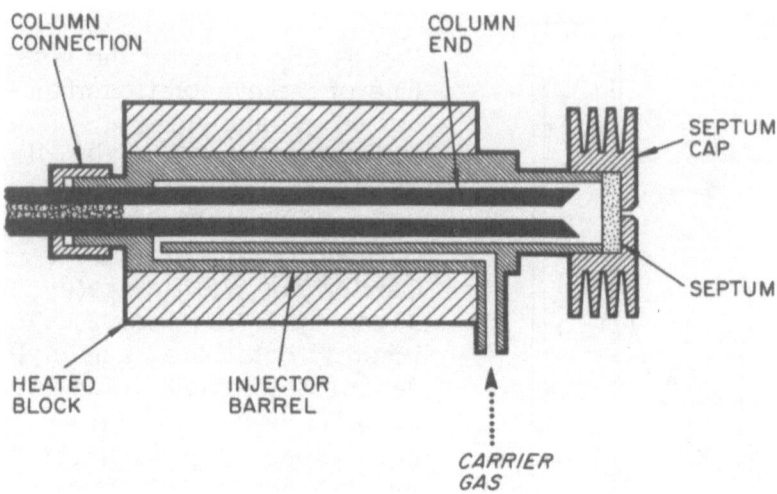

Fig.6. Simplified schematic of a sample introduction system with the extended front of a glass packed column serving as the evaporation chamber.

volume of the evaporation chamber must be selected according to sample requirements. Besides this, the carrier gas flow rate may also influence this selection: e.g., with flow rates below 10 ml/min, the use of narrow-bore liners (having small internal volume) is definitely recommended.

In the system shown in Fig.3 and discussed until now, the glass liner is an interchangeable part of the sample introduction system. Naturally, in the case of glass packed columns, the top of the column may also serve as this "liner". Such a system is shown in Fig.6; the top part of the glass column extends into the heated block thus serving as the evaporation chamber.

On-Column Sample Introduction

Sometimes systems such as the one shown in Fig.6 are also termed as on-column injection although they are not. True on-column injection must fulfill three basic criteria. First,

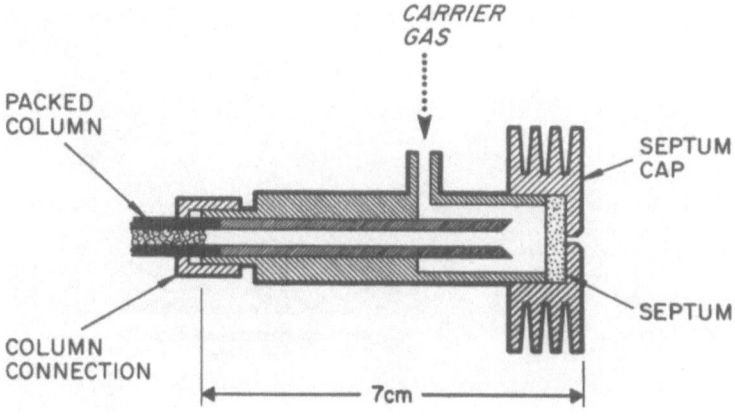

Fig.7. Simplified schematic of a truly on-column sample
introduction system.

the liquid sample should be deposited directly - without
evaporation - on the top of the column packing and, secondly, neither the syringe needle nor the liquid sample itself
should pass through a heated zone which is hotter than the
column prior to deposition on the head of the column.
Finally, the third criterion is that in the case of programmed
temperature applications, that part of the column where the
sample is deposited should be in the column oven. All these
criteria are fulfilled by the system shown in Fig.7. Here,
a Hamilton-syringe with a 7-cm needle deposits the sample
in the column, on the top of the packing material.

Split-Flow Sample Introduction Systems for
Open Tubular (Capillary) Columns

Earlier split sampling used in the case of capillary columns
was mentioned. In split sampling, a relatively large sample
is introduced, evaporated, and then mixed with the carrier
gas. Subsequently, this homogeneous mixture is split into
two highly unequal parts the smaller one being introduced
into the capillary column. Split sampling into such columns

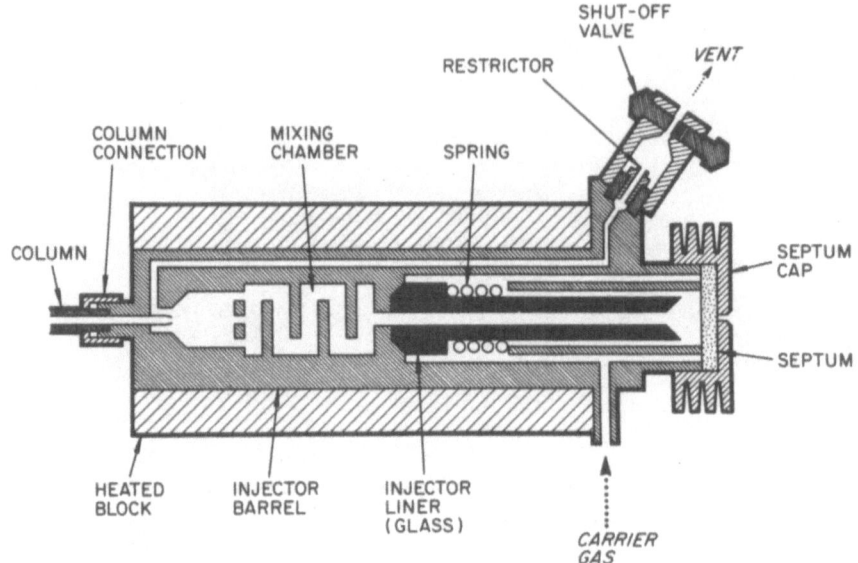

Fig.8. Simplified schematic of a split-flow sample introduction system for open tubular (capillary) columns.
"Classical" design with interchangeable vent restrictors.

has many special criteria; for details see the book chapter
of Condon and Ettre (1).

Split flow sampling systems have also undergone significant
developments during the last 15 years. Here, the construction of two advanced systems are shown.

The system shown in Fig.8 represents the most up-to-date
version of the so-called classical system. In this, the split
ratio is controlled by variable restrictors in the vent line.
A glass liner is also used here providing the evaporation
chamber.

In recent split systems designed to provide all-glass systems
for glass capillary columns (Fig.9), the glass liner advanced
to a kind of pre-column. The packing usually consists of
small glass beads providing the tortuous path needed for
homogeneous mixing; however, a part of this may consist of

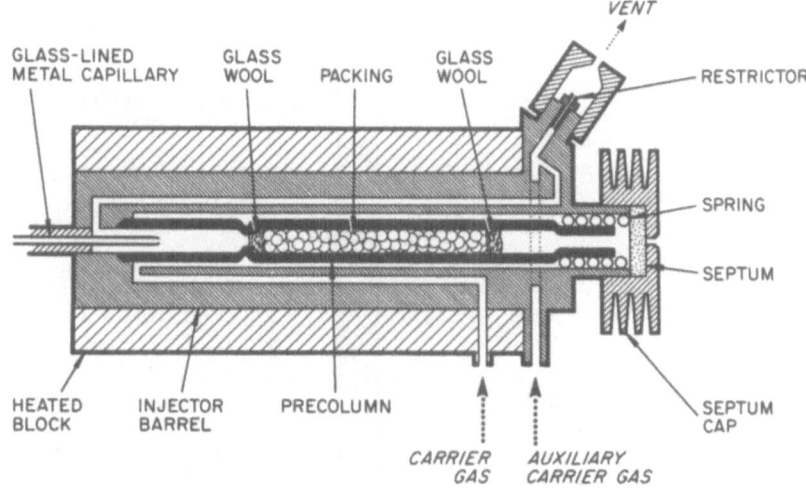

Fig.9. Simplified schematic of an all-glass split-flow sample introduction system for glass open tubular (capillary) columns.

an actual column packing recommended by some researchers as a way to prevent aerosol formation. In such a system, the split ratio is controlled by a secondary gas flow joining the vent line and providing the column inlet pressure. This is illustrated in the functional schematic of the system (see Fig.10). The design and performance of such systems are the subject of a number of recent publications (7-9).

SEPTUM BLEEDING AND POSSIBILITIES FOR ITS ELIMINATION

The second part of this paper concerns the problems associated with the common septum used in the sample introduction systems, and the possibilities of their elimination. These problems are usually summarized under the term of "septum bleeding". This expression refers to contamination introduced by the septum into the GC system which the detector cannot distinguish from "real" peaks. In temperature programmed runs the septum contamination may first condense at lower temperatures on the column and elute at a higher temperature creating real peaks; in fact, one may

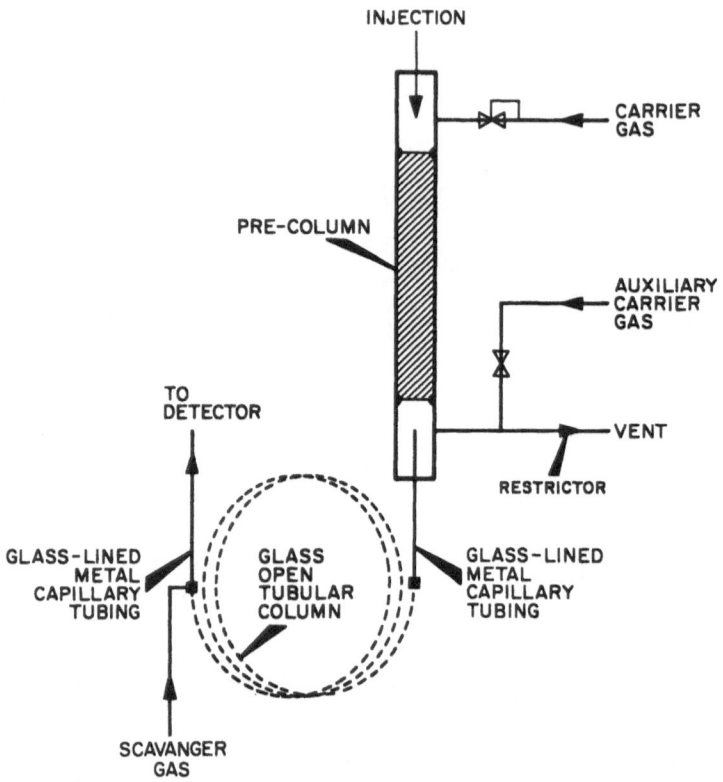

Fig. 10. The functional schematic of the all-glass split-flow sample introduction system for glass open tubular (capillary) columns.

produce in this way real "chromatograms" without the introduction of any sample. In isothermal analysis, the contamination is steady, and increases the background signal, thus
limiting detector linearity.

The contamination introduced by the septum may have three
origins. It may be originally part of the septum or adsorbed
on it during manufacturing or storage; it may come from
back-diffusion of the sample vapor in the injection block;
and, finally, it can be accumulated by the septum wiping
the syringe needle.

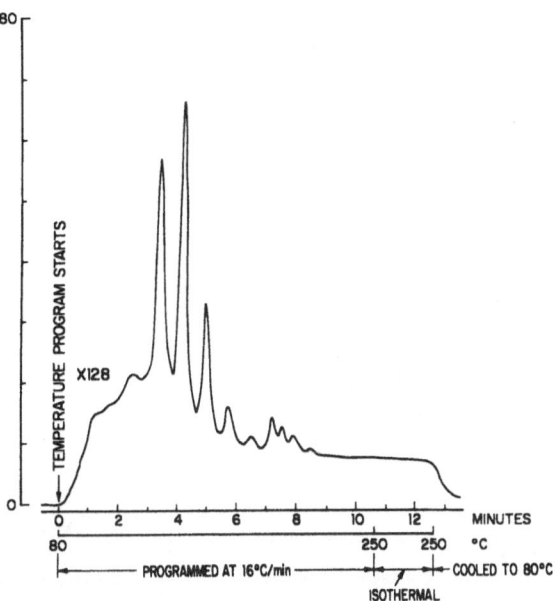

Fig.11. Septum bleeding after heating a new septum for 2
hours in the injector.
Column: 18 in. x 1/8 in.OD packed, 0.5 % SE-30 on
Chromosorb W HMDS 80/100 mesh. Carrier gas (He) flow
rate: 30 ml/min. Injector temperature: 300°C. FID. Amplifier
sensitivity at full scale and x1: 5 pA.

The extent of septum bleeding is demonstrated by Figs.11-
13. A brand new septum was placed in the injector cap,the
helium flow was adjusted to 30 ml/min, and the temperature
of the injector and oven was set to 300 and 80°C respectively. After two hours, the column was programmed through a
temperature cycle; the resulting recorder trace is shown in
Fig.11. The temperature program cycle was again repeated
21 hours after start (Fig.12) and then again, 45 hours after
start (Fig.13). It is clear that even after 45 hours very
distinct and real peaks were produced. It should be emphasized that this septum was never in contact with a sample!

Baking out the septum at elevated temperatures under constant purging or in vacuo helps somewhat to reduce the
"inherent bleeding", i.e., the amount of contaminants pres-

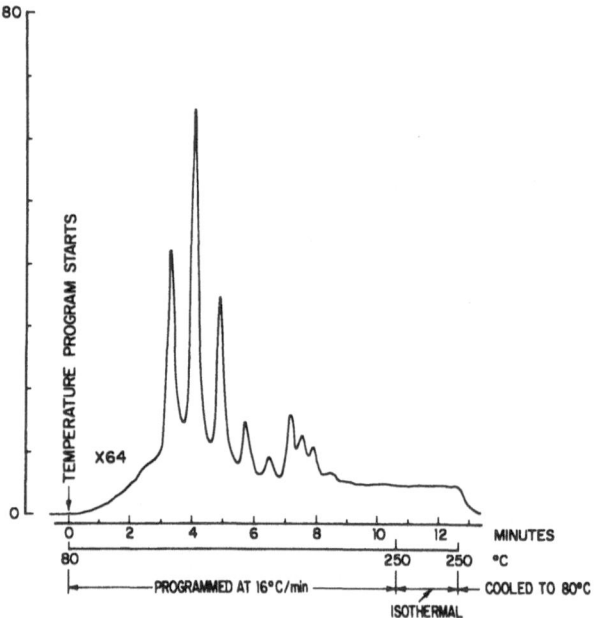

Fig.12. Septum bleeding after heating a new septum for 21 hours in the injector. Column and conditions as in Fig.11.

ent a priori in the septum, but will not completely eliminate it. And obviously, it cannot eliminate the collection of im- purities by the septum either upon storage or in a gas chrom- atograph. It should also be mentioned that baking a septum substantially reduces the puncturability - i.e., the number of times it can be pierced without creating a leak. Finally, it should also be emphasized that, according to our experi- ence, only the volatiles close to the surface area are re- moved in the baking process: during storage the volatiles present in the septum will slowly migrate to the surface and thus appear as bleeding when the septum is installed.

Another well publicized means of reducing the contaminants from a septum - both initially present or sorbed in the in- strument from back-diffused sample components - involves the use of septa having a thin Teflon* coating on their side facing the column. These septa work well until the first

* Trademark of E.I.DuPont de Nemours, Inc.

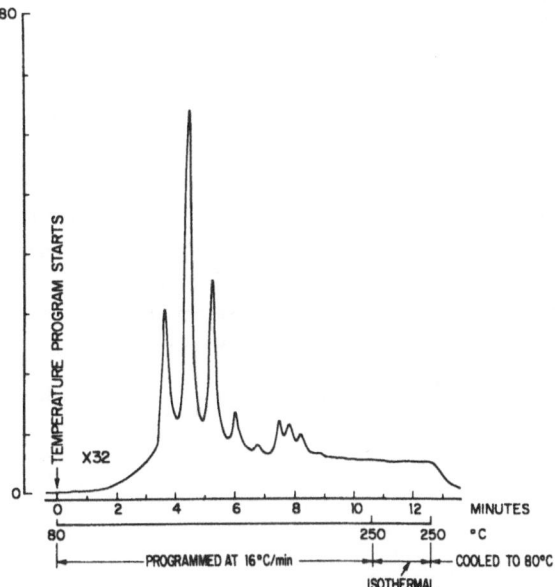

Fig.13. Septum bleeding after heating a new septum for 45 hours in the injector. Column and conditions as in Fig.11.

sample is injected: no bleeding leaves the septum proper. However, since Teflon is not self-sealing, a hole remains after the first puncture thus exposing the septum to the carrier gas. Furthermore, it is our experience that repetitive injections through such a septum will partially remove the Teflon coating from the septum creating a small pocket where the volatiles from the septum or back-diffused sample components can accumulate and, in turn, slowly diffuse into the carrier gas stream.

Back in 1967, we felt that a better way to limit the septum problems would be to separate the shielding functions from the septum. Accordingly, as shown in Fig.14, we placed a small bonnet made of Teflon and reinforced by stainless steel between the septum and the carrier gas flow path. This bonnet has a small - 0.03 in. - bore, just enough for the syringe needle. Thus, it has practically the same effect as the Teflon coating on the septum after the first injection but without the problems associated with the latter.

Table II. Legend to the Injector Schematics (Figs. 14, 17, 18 and 19.)

A	Injector barrel	J	Sliding valve
B	Injector sleeve	K	Stop pins
C	Injector liner (glass)	L	Seal (Teflon or aluminum)
D	Cooling fins	M	"O" ring (Teflon)
E	Septum	N	Rotating metal block
F	Septum knob	O	Screw
G	Injector seal (Teflon)	P	Handle
H	Shielding strip	Q	Spring
I	Stainless steel disk		

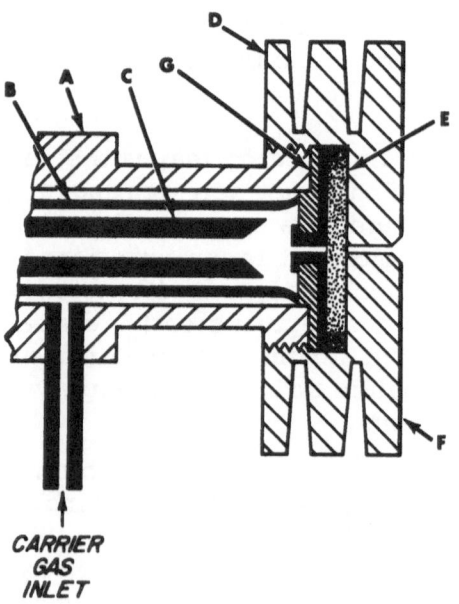

CARRIER
GAS
INLET

Fig.14. Simplified schematic of the front of an injection block, with a Teflon bonnet. For legend, see Table II.

These septum shields became standard part of Perkin-Elmer gas chromatographs for seven years.

The effect of this septum shield is demonstrated by two experimental series.

In the first case, 1 μl of a C_8-C_{18} even numbered n-paraffin sample was injected into a gas chromatograph without the bonnet and analyzed by programming the cclumn temperature from 80 to 250°C at 32°C/min. At the end of the program, the column was cooled to 80°C and then, the temperature cycle repeated but without introducing any new sample.The left-hand side of Fig.15 shows the recorder tracing during this cycle. It looks like a real chromatogram although, actually, no sample was introduced: the "peaks" seen are due to septum bleeding. At the end of the cycle when the column was at 250°C the injector cap with the septum was quickly removed, the bonnet installed and then the cap (with the septum) replaced. Now, the bonnet shields the septum from the carrier gas. After about 3.5 minutes purging the column was cooled to 80°C and the temperature program again repeated. As seen in Fig.15, the ghosting was reduced from off-scale to about 10% and, naturally, this also includes the normal column bleeding.

In the next example the bonnet was already installed prior to sample injection. The same normal paraffin sample was injected and the column temperature programmed from 80 to 250°C at 32°C/min as in the previous example; the resulting chromatogram - this is, of course, now a true chromatogram - is shown in the first part of Fig.16. After cooling the column to 80°C, the program was again repeated. Now, however, the bonnet shielded the septum preventing any traces of the sample from adhering to it. As a conclusion only a very small baseline disturbance was obtained which also includes the regular column bleeding. In other words, the bonnet had effectively shielded the septum.

Naturally,due to the small hole in the middle, the Teflon bonnet does not provide a complete shielding of the septum. Over the past couple of years, we have carried out a very extensive development work in this respect and designed a

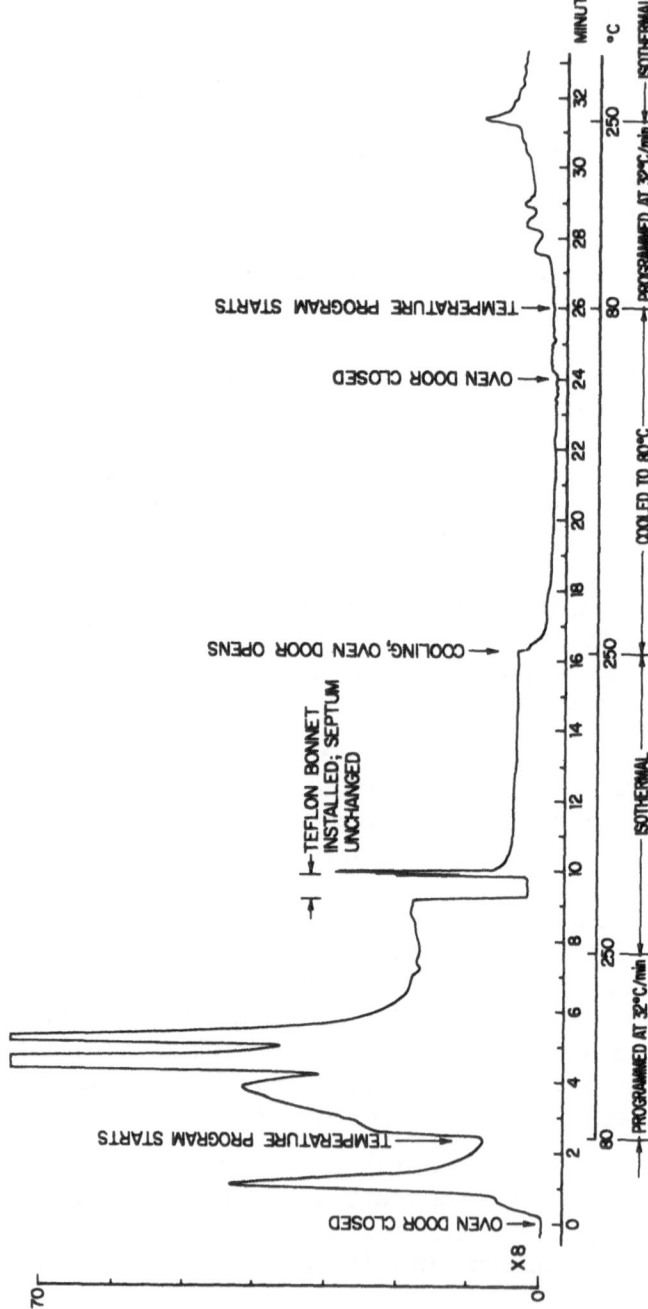

Fig.15. Background with and without the Teflon bonnet.
Column: 18 in. x 1/8 in. OD packed, 0.5 % SE-30 on Chromosorb W HMDS 80/100 mesh.
Carrier gas (He) flow rate: 30 ml/min. Injector temperature: 300°C. FID. Amplifier sensitivity at full scale and x 1: 5 pA.

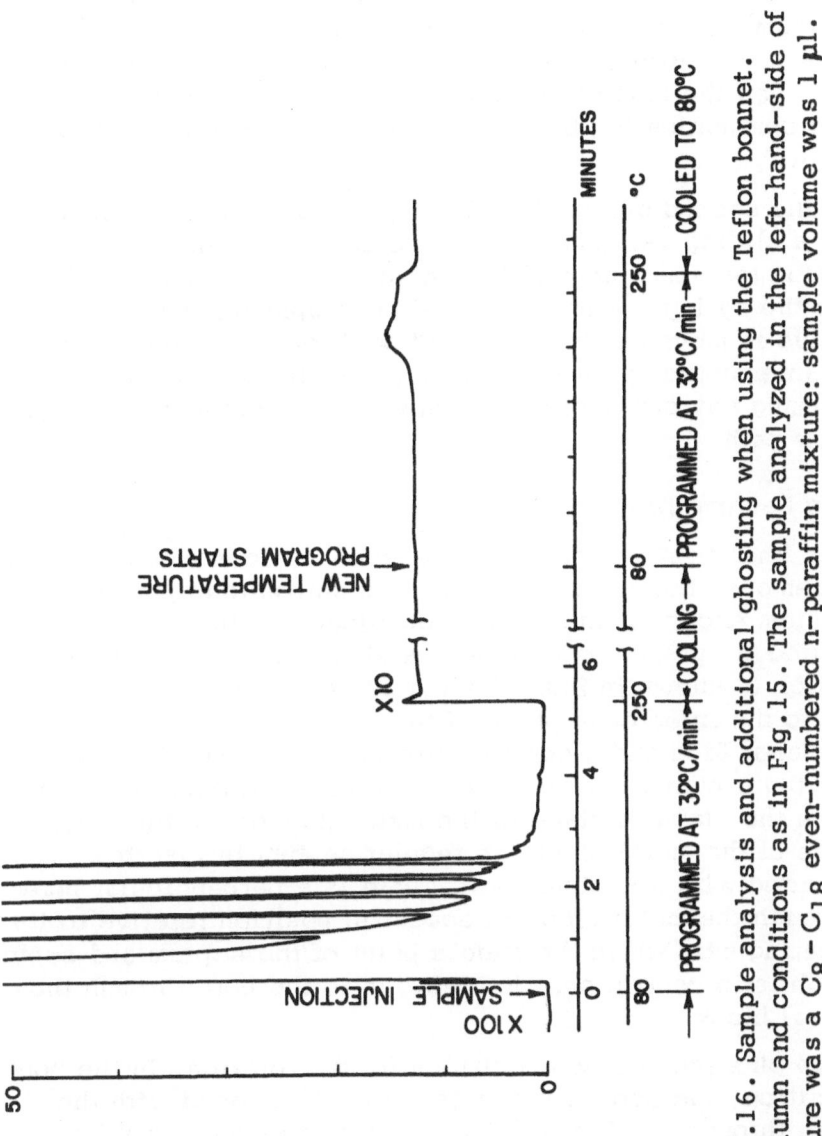

Fig.16. Sample analysis and additional ghosting when using the Teflon bonnet. Column and conditions as in Fig.15. The sample analyzed in the left–hand–side of the figure was a $C_8 - C_{18}$ even–numbered n–paraffin mixture: sample volume was 1 μl.

number of systems. Two of these are illustrated here because they represent interim stages of the final design. In one of these (Fig.17) a Teflon strip is placed in front of the septum. After an injection, the strip is immediately advanced to a new position so that there is always an unbroken layer between the septum and the carrier gas thus reducing to zero the communication between the septum and the gas flow path.

In the second design (Fig.18) the Teflon strip is replaced by a sliding valve having two positions. In the "open" position the small hole in the valve is aligned with the channel for the syringe. Immediately after sample injection, the valve is moved to the "closed" position where the channel is interrupted by an inert, non-permeable shield thus preventing any communication between the septum and the gas flow path.

The Septum Swinger

Our final design represents an advanced form of these ideas. It removes the septum from the gas chromatographic injection system except at the moment of injection. This is the so-called septum swinger, a simplified schematic of which is shown in Figs. 19 and 20. A metal flange is attached to the injector in place of the regular septum cap, having a planar face and a centric through bore. A metal block having a planar surface is disposed in confronting rotation with the planar surface of the metal flange. On the other side of the metal block the regular septum is inserted and secured with a septum knob. There is a perpendicular bore through the metal block in eccentric position relative to its rotating axis while the middle point of the septum and septum knob is in centric position relative to the bore through the metal block.

Stop pins provide two positions for the rotation. In the "on" position, the bore of the metal block is aligned with the bore through the flange. Thus, a syringe needle can be inserted through the septum into the inside of the injector. Immediately after withdrawal of the syringe needle, the block is turned to the "off" position in which the bore in the

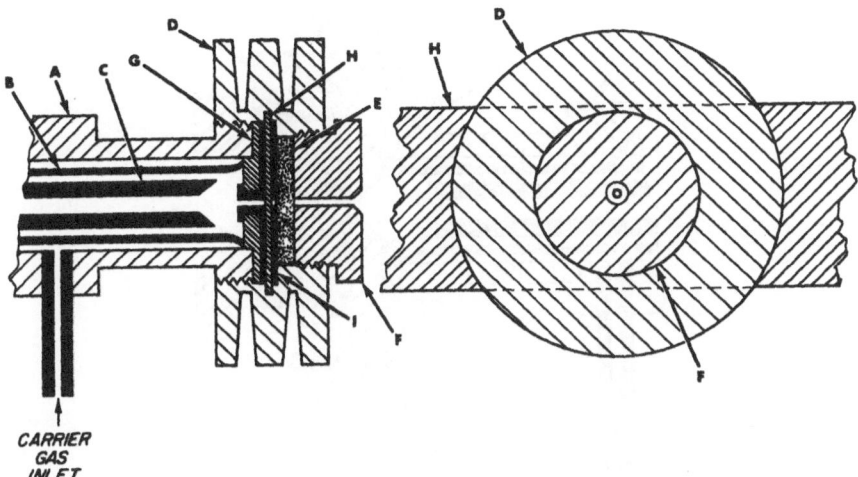

CARRIER
GAS
INLET

Fig.17. Simplified schematic of an injector cap with a movable Teflon strip. For legend, see Table II.

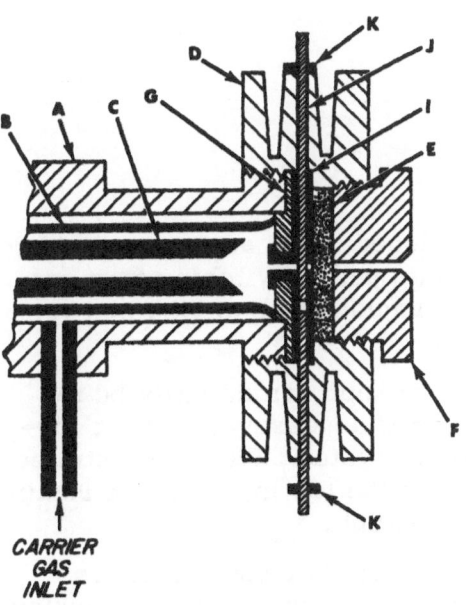

CARRIER
GAS
INLET

Fig.18. Simplified schematic of an injector cap with a shielding two-position sliding valve. For legend, see Table II.

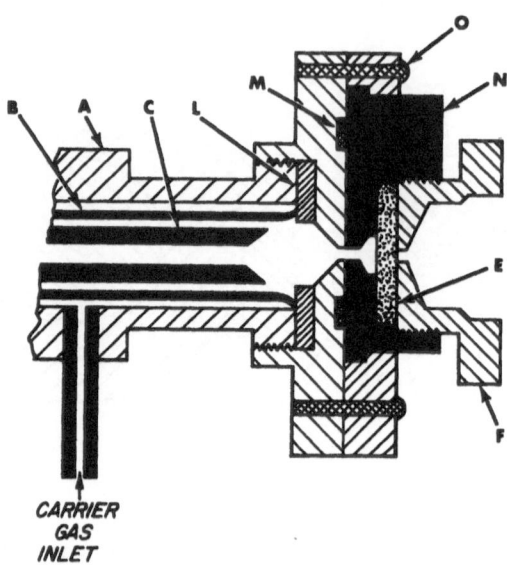

CARRIER
GAS
INLET

Fig.19. Simplified cross-sectional schematic of the Septum
Swinger attachment. The unit is in "on" (open) position.
For legend, see Table II.

flange now faces a highly polished metal surface, providing
a complete seal of the inner gas path. Switching from one
position to the other is very simple and is guided by two
stop pins.

In the "off" position the septum knob can actually be removed and the septum replaced if necessary, without dis-
turbing the gas chromatographic system. Fig.21 illustrates
the Septum Swinger installed on the top injector of a Perkin-
Elmer Model 910 Gas Chromatograph.

The complete shielding effect of the Septum Swinger is
illustrated by a few consecutive measurements. The system
consisted of the Model 910 Gas Chromatograph equipped
with two injectors with a separate column in each channel,

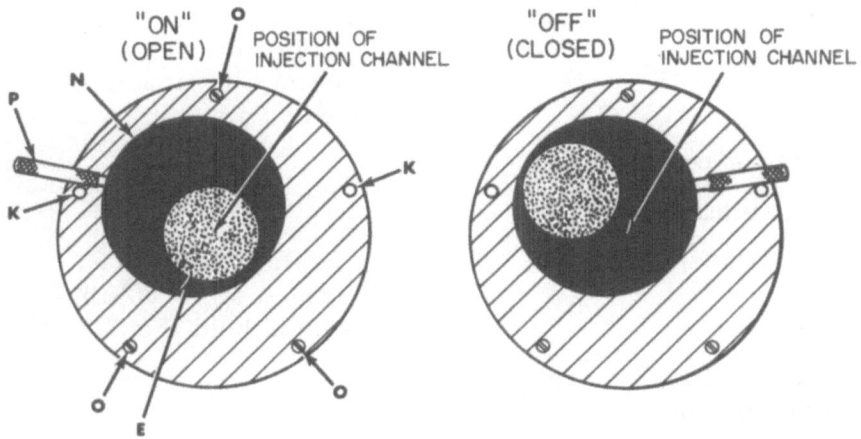

Fig.20. Front view of the Septum Swinger attachment indicating the two positions. The septum knob is removed.
For legend, see Table II.

Fig.21. The Septum Swinger
installed on the top injector
of a Perkin-Elmer Model 910
Gas Chromatograph.

The unit is in "on" (open)
position.

Table III. Analytical Conditions for the Chromatograms
Shown in Figures 22-25.

Column	two 18 in. x 1/8 in. OD packed 0.5 % SE-30 on Chromosorb G HP 80/100 mesh
Injector temperature	300 °C
FID temperature	300 °C
Column temperature	5 minutes isothermal at 80°C then programmed at 25°C/min up to 250°C and hold there
Carrier gas (He) flow rate in each column	30 ml/min
Recorder	Leeds & Northrup Model 540 dual pen; 1 mV full scale sensitivity at x1; 10 mm/min chart speed

having a separate flame ionization detector at their end.
Detector responses were recorded on a dual-pen recorder.
The columns employed had been in use for some time at
high temperatures and thus, showed practically no liquid
phase bleeding.

Two new silicone rubber septa were taken from the same
batch and installed in the regular septum caps, without any
shielding. No sample was injected during the entire experimental series: what is shown here is pure septum
bleeding. After instrument stabilization, the columns were
programmed up to 250°C (for analytical conditions see
Table III). Fig.22 shows the detector responses; as can be
seen, the septum bleed was almost identical in each channel.
This could be expected since both septa were of the same
batch, having the same history, and flow rates and temperature cycles were identical in both channels.

Next, the septum cap was removed from injector B and replaced by a Teflon plug. After a few minutes purge, the

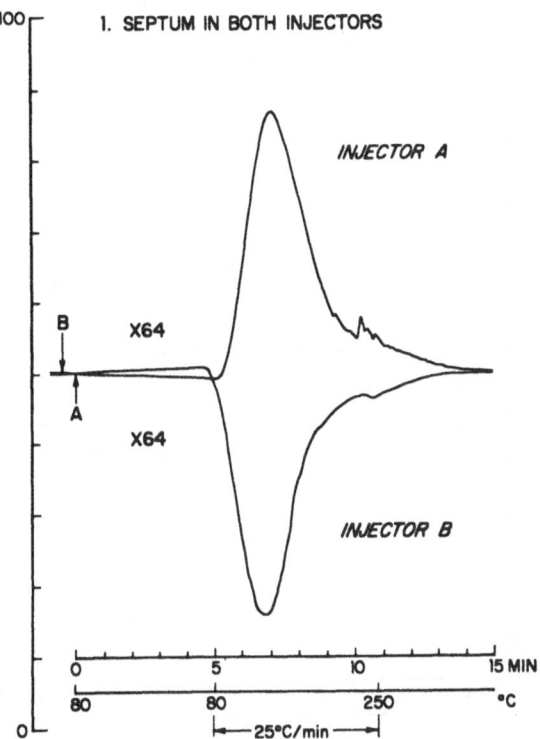

Fig.22. Investigations of septum bleed, I.: Septum in both injectors. For conditions see Table III.

columns were cooled to 80°C and then programmed through the cycle. As shown in Fig.23 the septum in channel A gave an almost identical bleeding as earlier but, of course, there was no bleeding in channel B since there was no septum there.

Subsequently, the septum originally present in channel B was placed in a Septum Swinger which was then installed in place of the Teflon plug. After system equilibration, the columns were cooled to 80°C and the Septum Swinger was switched for 10 seconds to the "open" position to simulate the situation during a sample injection. Then the Septum Swinger was closed and the columns programmed through

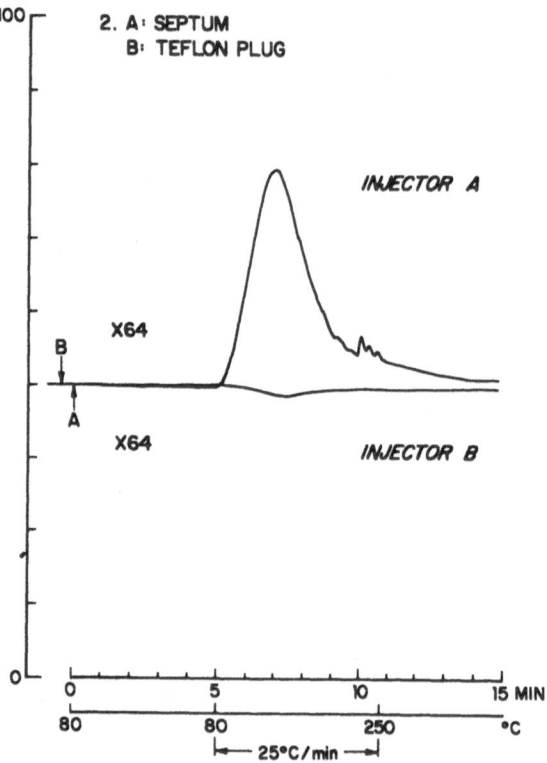

Fig.23. Investigations of septum bleed, II: Teflon plug in injector B. For conditions see Table III.

the temperature cycle. Fig.24 shows the resulting recorder trace: as seen, no septum bleed could be observed from the septum in the Septum Swinger.

Finally, to prove that the septum still had all the accumulated volatile material on its surface, it was removed from the Septum Swinger, placed into the original septum cap and the Septum Swinger was replaced by this cap. Again, the columns were first cooled to 80°C and then the temperature cycle repeated. Fig.25 shows the results: indeed, a high bleeding was obtained from this septum which showed no bleeding when used in the Septum Swinger.

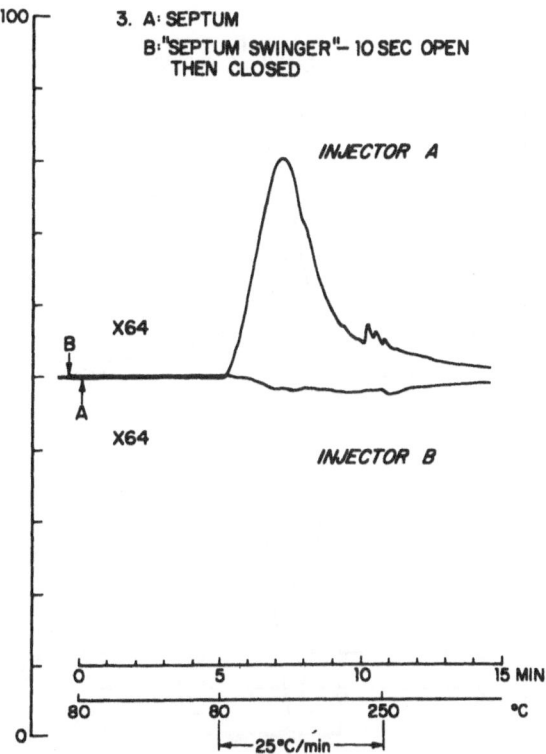

Fig.24. Investigations of septum bleed, III.: Septum Swinger on injector B. For conditions see Table III.

CONCLUSIONS

The evolution of present-day sample inlet systems was illustrated and the design of modern, up-to-date systems discussed. None of these can be termed as "universal": each must have an optimum construction for the particular column and sample type and size, and carrier gas flow rate. Only by the selection of the proper – and optimum – sample introduction system can the true efficiency of the gas chromatographic column be fully utilized.

Septum bleed may be detrimental in creating artifact peaks and/or increasing detector background. Its complete elim-

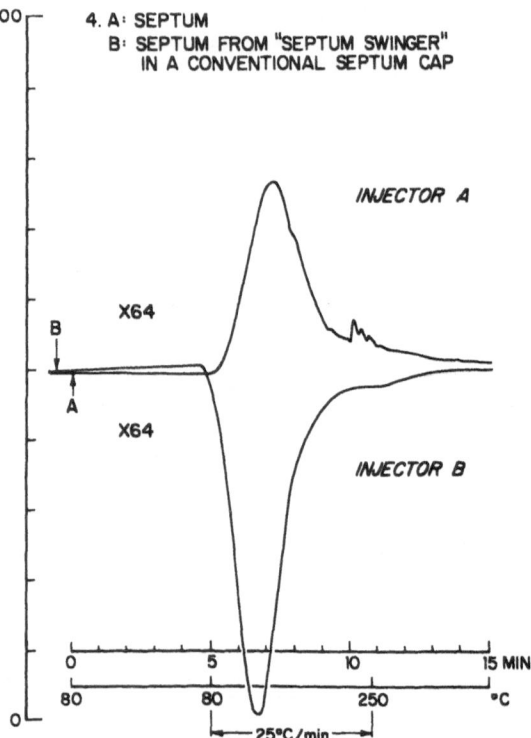

Fig.25. Investigations of septum bleed, IV,: Septum from Septum Swinger placed in a conventional septum cap and reinstalled in injector B. For conditions see Table III.

ination is possible with the Septum Swinger which, while maintaining the convenience of injection with a syringe through a septum, removes the septum from contact with the carrier gas.

ACKNOWLEDGEMENTS

The development of the various sample introduction systems and the Septum Swinger was carried out in cooperation with a number of our colleagues from whom we would like to particularly mention R. D. Condon, H. D. Downs,

M. J. Hartigan, and M. W. Redmond. We would also
like to express our appreciation to B. Giordano who
carried out part of the evaluation work of the Septum Swinger

REFERENCES

1   R.D.Condon and L.S.Ettre: "Liquid Sample Processing".
    In Instrumentation in Gas Chromatography, J.Krugers,
    ed., Centrex Publishing Co., Eindhoven, 1968; pp.
    87-109.

2   J.E.Purcell, H.D.Downs and L.S.Ettre, Chromatographia
    8, 605 (1975).

3   A.T.James and A.J.P.Martin, Biochem. J. 50, 679 (1952).

4   N.H.Ray, J.Appl.Chem. 4, 21 (1954).

5   H.H.Hausdorff: "Vapour Fractometry (Gas Chromatog-
    raphy)". In Vapour Phase Chromatography (1956 London
    Symp.), D.H.Desty, ed., Butterworths, London, 1957;
    pp. 377-387.

6   M.J.Hartigan and J.E.Purcell, Chromatogr.Newsl. 3
    (1), 1-22 (1974).

7   A.L.German and E.C.Horning, J.Chromatogr.Sci. 11,
    76 (1973).

8   A.L.German and E.C.Horning, Anal. Letters 5 (9),
    619 (1972).

9   M.J.Hartigan and L.S.Ettre, J.Chromatogr. (to be
    published).

GAS-SOLID CHROMATOGRAPHY: APPLICATION, UNIQUENESS AND VERSATILITY

Robert L. Grob

Villanova University

Villanova, Pa. 19085

INTRODUCTION

Gas Solid Chromatography (GSC) has been applied to the separation of mixtures of hydrogen isotopes, isomers, proteins and viruses. This interest and development has been due to:
1) Control of Homogeneity and specificity of molecular sorbents, i.e., synthesis of absorbents or modification of their surfaces.
2) Expansion of temperature range of chromatographic columns (500°C).
3) Use of strongly adsorbable carrier gases which can be used at high pressures.
4) Control of the nature and degree of porosity of adsorbents.
5) Development of non-swelling molecular and macromolecular sieves.
6) Importance as a tool in physicochemical investigations of solid surfaces, studies of isotherms, determination of heat capacity of adsorption systems and determination of heats and entropies of adsorption.

GSC really began to develop in the years 1953-1959. Historically we could go back to the 1920-30 period where Berl (1,2,3) studied the sorption and desorption of gases at low temperatures. Although Berl's work was not exactly Gas-Solid Chromatography, it did lead to the work of Peters (4,5,6) who showed the effect of pressure variation on the sorption processes. Thus, we could say that Peters' work was the predecessor of displacement chromatography. Euchen

151

and Knick (7) studied the effect of increased temperature
to separate adsorbed substances. In their publication they
discussed the theoretical and empirical aspects of adsorption of gases. Papers by Henjes (8) and Ferber and Luther
(9) followed.

The separation of gas and/or vapor mixtures by the
technique of desorption from an adsorbent was studied by
two methods:
1)   A movable heating element was passed along a
column. This method had been utilized, initially, by
Turner (10) and then applied to an instrument marketed by
Burrell Company of Pittsburgh, Pa. (11).
2)   An inert gas was passed through the column containing the adsorbent. This would be very similar to the
transport of a liquid down a column in liquid chromatography.
The inert gas caused the desorption of each of the adsorbed
species in some particular order. This desorption process
resulted in an "elution" curve as shown in Figure 1.
Claesson (12,13) improved upon this technique by using a
carrier gas containing a fixed concentration of a substance
more strongly adsorbed than any of the sample components.
This resulted in a displacement curve (Figure 2). In the
work of Claesson the step height was constant and characteristic for a specific component, i.e., it corresponds to the
retention time, t_R. The step length was found to be proportional to concentration of component in the mixture,
i.e., it corresponds to the peak height. Nitrogen was used
as the carrier gas and his analyzer was a thermal conduc-
tivity cell. The accuracy of the measurements (analysis)
was found to depend upon several factors:
1) Column Size Component fronts increased with
smaller diameter tubing. For a given column diameter, short
column lengths gave maximum separation. In order to work
with larger sample sizes, and still have sharp fronts, the
column was built in sections. Each column section was
shorter than the one which preceded it in the system.
Figure 3 illustrates this effect for ethyl acetate on activated charcoal.
2) Adsorbent Charcoal was used for all studies.
Sieving and drying temperatures were varied. Improvement
in component fronts was accomplished by saturating packing
with a liquid which was subsequently removed by heating
column in a higher boiling liquid (surface modification) or
by coating adsorbent with a substance not displaced or

appreciably eluted (gas-liquid chromatography).

3) <u>Particle Size</u> Decreasing the particle size (in-creased homogeneity) made component fronts sharper. Going to very small particles caused large pressure drops across the column and increased the probability of dust, which produced irregular and diffuse fronts.

4) <u>Flow Rate</u> Decreasing the velocity of the carrier gas improved the sharpness of the fronts. A minimum velo-city was reached where further decrease was not effective for improvement of fronts. Fast velocities, >100 cm^3/min, caused irregularities in the output of the detector. This is analogous to the effects seen by the van Deemter plot. An interesting point is discussed and substantiated by the data of Claesson. The product of the step height and spe-cific length was a constant, corresponding to peak area in elution work and attesting to linearity of thermal con-ductivity detectors.

Early work showed that substances with similar boiling points resulted in poor resolution; in fact, such sub-stances had almost the same step height. To circumvent such problems, chemical and physical properties were utilized to improve the separation. An example of this was the sepa-ration of butane, butene and propane (12). Butane and bu-tene came off together (i.e., one step). Passing the col-umn effluent over sulfuric acid (conc.) soaked pumice re-moved the butene. This is illustrated in Figure 4.

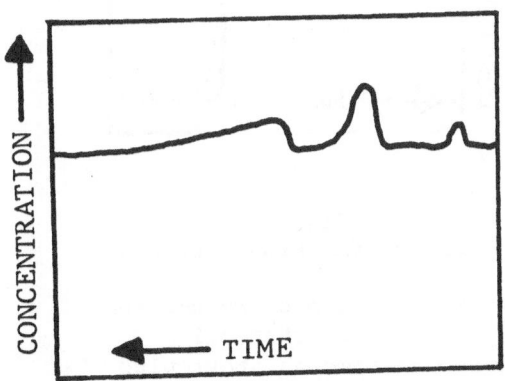

Figure 1

Elution Curve

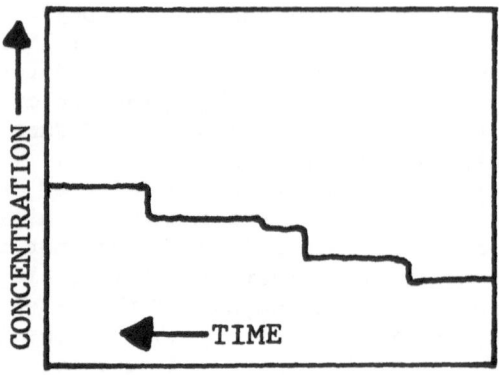

Figure 2
Displacement Analyses

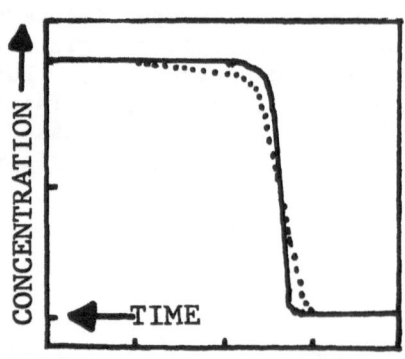

Figure 3
Fronts for Ethyl Acetate

- - - - 9mm o.d. column and
        $4cm^3$ charcoal
———— same as above plus
        2.5mm o.d. column
        and $0.5cm^3$ charcoal

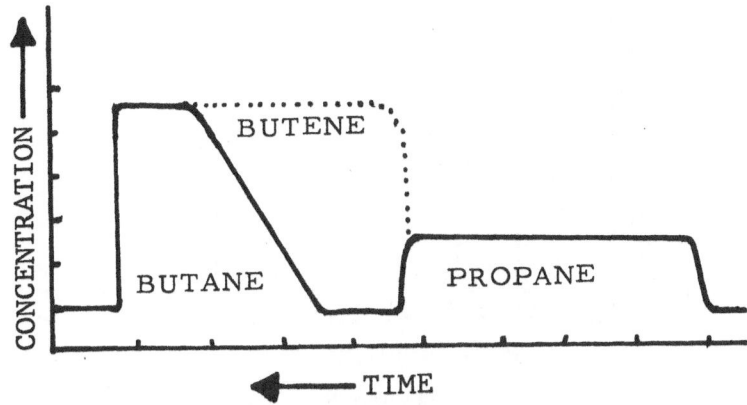

Figure 4
Separation of butane, butene and propane

These initial studies also showed that the amount of one substance was independent of amount of other components. Thus, it was realized that small percentages of one component could be estimated in presence of large amounts of other components.

Persons interested in additional historical and theoretical readings pertaining to GSC should read papers by Cremer (14,15), Janak (16,17), Patton (18), Turkeltaub (19) and Cremer and Roselius (20).

It was generally believed that GSC was limited in its versatility for the analysis of gases or conveniently volatilized compounds. As we shall show it does have considerable technological promise.

Some of the advantages of GSC are:
1) Higher column efficiencies can be realized because of the absence of a liquid-phase contribution to band spreading.
2) High column temperature limits (500°C).
3) Stable solid surfaces which do not undergo oxidation or other chemical reactions.
4) Selective separations; large specificity for geometrical isomers.
5) Absence of "bleed" problems that occur with liquid

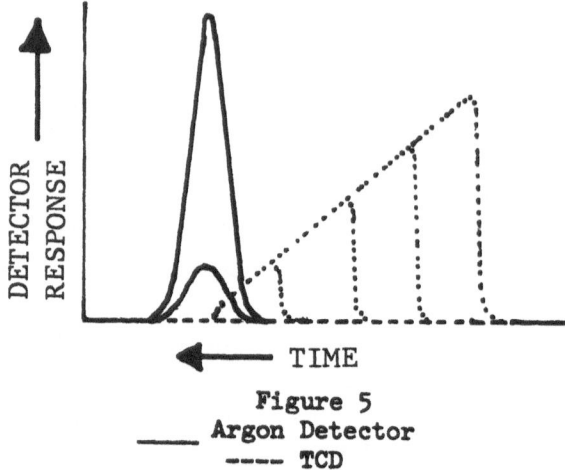

Figure 5
Argon Detector
---- TCD

phases and cause contamination and background noise for the
more sensitive detectors.

The importance of a sensitive detector, in gas-solid
chromatography, is illustrated in Figure 5. The katharo-
meter detector requires larger sample sizes; thus, requiring the use of the non-linear region of the adsorption
isotherm. This results in skewed peaks. By contrast, the
argon detector requires smaller sample sizes and thus utilizes the linear region and furnishes gaussion shaped peaks.
Giddings (21) has stated that tailing may also occur when
the adsorption isotherm is linear. He attributed this to
a kinetic phenomenon arising from the long average desorption time of some adsorbates from "active" sites on the adsorbent. It was also Giddings (22) who laid the basis for
the theoretical treatment of efficiency and resolution in
GSC. Peak broadening, at high carrier gas flows may be
controlled by the kinetics of mass-transfers. Additional
advantages of this technique are:
      6) The kinetics of adsorption-desorption are much
faster than the corresponding process of diffusion in the
liquid phase. This permits smaller HETP values in GSC than
in Gas-Liquid Chromatography (GLC). Grubner (23) discussed
the statistical moments theory of GSC and gave relationships
between the centered peak moments and the various parameters
of the adsorbent.

7) GSC is an important technique in the physicochemical studies of solid surfaces, the study of isotherms and heats and entrophies of adsorption.

8) Catalysis studies; physicochemical properties, catalyst evaluation, kinetics of catalytic reactions and study of catalytic reactions.

PHYSICAL ADSORPTION OF GASES AT SOLID SURFACES

All processes, at the gas—solid interface, whether it be chemisorption, catalysis, dissolution or a heterogeneous chemical reaction, includes as part of its mechanism, physical adsorption of a vapor or gas on the solid surface. An intriguing high-light of adsorption theory is the calculation of adsorption energies from electrical, magnetic and the geometrical properties of the adsorbate and the solid adsorbent. It is doubtful that the surface layer of crystal can be considered as a surface formed as a result of some geometrical plane intersecting an ideal crystal.

No specific adsorption forces exist in nature; so one should regard the problem of calculating adsorption energies as a particular aspect of the area of intermolecular interactions. One could speculate that the intermediate distances between sorbed molecules should be shorter than those encountered in studying real gases but longer than those distances characteristic of chemical interactions. The current theory of intermolecular interactions does not allow one to express the wave function in a general form; thus, different types of interaction are regarded as being independent. Energies of adsorption for non-polar molecules on homopolar crystals limits one to the dispersion forces of attraction and repulsion; whereas adsorption on heteropolar crystals with ionic lattices, electrostatic interactions are considered. Basic work, in this regard, has been carried out by Barrer(24), who calculated the energy of adsorption on graphite and Orr(25) on ionic crystals of KCl and CsI. Both studied the interactions of simple monoatomic and diatomic molecules. Studies of this nature were expanded by Kiselev and workers(26-31) for the calculations of adsorption energies of simple molecules and polyatomic, non-polar hydrocarbon molecules on the basal plane of graphite. Their results were compared with very precise calorimetric measurements of adsorption heats and with sorption isotherms of the respective gases or vapors on graphitized

carbon blacks. Kiselev and Poshkus(32-34) studied the energy of interaction of hydrocarbon molecules on magnesia
and magnesium hydroxide solids.

We can estimate the theoretical value of interaction
energy between an isolated molecule and the surface of a
solid in the following manner. Consider that the gas consists of simple spherical molecules located in a field of
infinite adsorbent. Also, that the temperature is high
enough to neglect molecular interactions in the gas phase.
The gaseous state may be described by an equation of state
of a real gas with virial coefficients(35-37). Introduce,
N, gas molecules at a temperature, T, and pressure, P,
into the container which holds the adsorbent. The adsorbent has a uniform surface ("dead" volume, V_g), so one can
determine the so-called "apparent volume", $V_a(=NkT/P)$. If
we designate V_o as the limiting value of V_a as P goes to
zero, we can write(38):

$$1/V_a = 1/V_o + PV_a \, C/kTV_o^3 \qquad\qquad (1)$$

$$V_o = V_g + \int V_g \quad \exp(-E/kT)dV \qquad\qquad (2)$$

where E = potential energy of gas molecules in adsorbent
field. The dependence of $1/V_a$ on PV_a will be a straight
line having an intercept $1/V_o$. The value of E then may be
calculated by integrating the second equation above. Equations obtained by this virial coefficient method make it
possible to estimate interaction energy and the specific
surface of the adsorbent. Heats of adsorption may be obtained by direct calorimetric methods or the determinations
of temperature dependence of adsorption. This is an area
where GSC can be used advantageously(39). Entropy changes,
which accompany adsorption processes, can be determined
from adsorption isotherms, provided one knows the heat of
adsorption in conjunction with the free energy changes.
Even though some solids have energetically inhomogeneous
surfaces, this has been handled very nicely by some investigators. The low coverage, in the case of porous
zeolite crystals, has been found to result from the gradient
of the electrostatic field, which contains the electrically
assymetric sorbate molecules possessing dipole or quadrupole
moments(40-41). Detailed studies have been made for carbon
blacks (low temperature treatment), having non-uniform surfaces, and carbon blacks graphitized at high temperatures
(42-43).

The mechanism of physical and chemical adsorption is of no consequence, from the thermodynamic viewpoint, when one regards the adsorption equilibrium as an equilibrium between the adsorbate in the gas phase with the adsorption complexes on the surface of the adsorbent. Kiselev(44-45) derived an equation for the adsorption isotherm, using a quasi-chemical equilibrium method, which took into account lateral interactions (from single and multiple complex formation). This was later extended to include sorbed molecules normal to the surface and tangential interactions in the first layer.

An interesting argument has been advanced(46) regarding the meaning of a Langmuir Isotherm. It is generally accepted that the Langmuir equation requires the adsorption enthalpy, ΔH, and the entropy, ΔS, to be independent of coverage, Θ. These authors found that the Langmuir equation may be obeyed when ΔH and ΔS are dependent upon Θ, provided the following condition is fulfilled:

$$(\Delta H/T) - \Delta S = \text{Constant} \qquad (3)$$

This has been found to be the case when iodine is adsorbed on Linde 13X molecular sieve (synthetic zeolite).

PROPERTIES OF ADSORBENTS–ADSORPTION VARIABLES

Adsorption properties of sorbents, with respect to a particular sorbate generally depend on three factors:
1) The nature of solid surface, i.e., chemical composition of surface layer and its crystal structure.
2) Specific surface area.
3) The nature of the solid's porous structure.

Much of the research into the nature of sorbent surfaces can be divided into:
a) Analysis of the chemical and crystallographic nature of surfaces.
b) Modifications of these surfaces by "chemical" or "physical" treatment.

The chemical analyses of sorbent surfaces is of importance because IR spectroscopy is difficult due to scattering and continuous absorption and results of elec-

tronic paramagnetic resonance measurements are not easily
interpreted, unambiguously.

The bulk of the work in the area of carbonaceous sorbents has been in the direction of surface modifications of
graphitized carbon blacks, controlled oxidation and bromi-
nation. Silica and other oxide sorbents have been modified
by hydration and dehydration, as well as such chemical reactions like etherification, fluorination and methylation.

Although a large amount of work has been expended in
the direction of specific surface area measurements, the
main source of adsorption phenomena remains to be the
analysis of adsorption isotherms. Methods used for determination of specific surface area have been directed. into
the technique of chromatography(47-48), radiometric analyses (49-50), heats of wetting(51-52), gas filtration
(53-55) and X-ray small angle scattering(56-57).

Freeman(58), by means of adsorption isotherms and equations with virial coefficients, was able to determine the
product of surface area and thickness of sorbed layer. By
proper assumption of the magnitude of the sorbed layer one
can calculate the surface area of the solid.

One is able to determine sorbent porous structure from
the analysis of sorption isotherms. In order to do this,
it is necessary to determine the sorbent structural type
by some independent method other than adsorption measurements alone(59). Among these other methods are electron
microscopy, X-ray small angle scattering, mercury poro-
metry and use of luminescent dyestuffs (pore sizes 10-1000
A).

Measurements of the sorbed material have been performed by such physical methods as, IR spectroscopy, magnetic susceptibility, paramagnetic resonance and determination of dielectric properties. Spectral measurements
are very important in chemisorption studies; strong bands
between gas molecules and solid surfaces result in sharp
changes in the spectrum. Physical adsorption can also be
followed by spectral changes but more information is obtained if the spectral data is compared with the isotherms
and heats of adsorption. Crawford(60) has reviewed this
aspect of reactions at gas-solid interfaces.

Interactions of methyl bromide with porous glass surfaces has been studied and it was decided that the methyl
bromide molecules lose some of their rotational degrees of
freedom. This is very easily followed by comparing
spectrum of gaseous methyl bromide to that of the sorbed
methyl bromide. It was interpreted to imply formation of
hydrogen bonds between the bromine atoms and surface
hydroxy groups(61).

MOLECULAR INTERACTIONS

The adsorption properties and selectivities of gas-
solid column packings are influenced the most by the adsorbent surface chemistry. Geometric heterogeneity of the
surfaces and pores can be largely reduced by using crystalline porous and non-porous adsorbents, amorphous adsorbents
with sufficiently wide pores or by modifying their surfaces
chemically. Theoretical treatment of the interactions of
molecules of a gas mixture, in contact with a homogeneous
surface, is more easily performed than when the molecules
are dissolved in the bulk of a liquid film. This is apparent if one considers molecules in a liquid as compared
to the molecules adsorbed on a surface. In the former environment the molecules are mobile and surrounded on all
sides by other molecules; whereas in the latter case when
adsorbed on a smooth solid surface the molecules interact
primarily with nearest force centers of the solid (these
centers are fixed). There is no general expression for the
potential of short-range intermolecular interaction. Contributions of various interactions, such as dispersion,
electrostatic, repulsive and chemical interactions come
into play during adsorption. The interactions to which we
are referring are molecular not chemical, i.e., molecular
physical adsorption not chemisorption.

Kiselev(62-64) has classified molecules and adsorbents
according to their specific and non-specific nature and intermolecular interactions. A summary of these classifications would aid in pointing up the various parameters one
can manipulate to study gas-solid chromatographic separations

Group A Molecules

Molecules with a spherically symmetrical electron

shell, eg., noble gases and saturated hydrocarbons having
only sigma bonds between carbon atoms (no elongated orbi-
tals-no locally concentrated electron density on periph-
eries). These molecules interact nonspecifically, through
universal dispersion forces resulting from concordant electronic motion in the interacting molecules.

Group B Molecules

Molecules with concentrated electron density (negative
charge), eg., unsaturated and aromatic hydrocarbons and all
molecules with pi electron bonds (N_2, H_2O, ROH, ROR, RCOR,
NH_3, NH_2R, NHR_2, NR_3, RSH, RCN).

Group C Molecules

Molecules with locally concentrated positive charges
within small radius linkages. There should not be adjacent
linkages with concentrated electron density on the periphery, such as -OH or =NH groups. Examples of this group
are organometallic compounds. These would interact specifically with Group B molecules but non-specifically
with Group A molecules.

Group D Molecules

Molecules possessing adjacent links of small radius
with positive charge in one and electron density concentrated on periphery of other. Types would be molecules
with -OH and =NH functional groups, eg., H_2O, ROH, primary
and secondary amines. Molecules of this group may interact specifically with group B and C molecules and each
other; they will interact non-specifically with group A
molecules.

Analogously, the adsorbents may be classified according to their ability to interact with molecules:
  1) Nonspecific Adsorbents. Adsorbents of this type
carry no functional groups and no exchange ions, eg., carbon blacks, boron nitride(BN) and saturated hydrocarbons
(especially polymeric types as polyethylene).
  2) Specific Adsorbents with Concentrated Positive

<u>Surface Charges</u>   This type have acidic hydroxyl groups on surface, eg., hydroxylated acid oxides such as silica; also adsorbents with aprotic acid centers or small radius cations, eg., zeolites. In the latter example the positive charge is concentrated in the exchange cations and the negative charge is equalized throughout the internal bonds of the aluminate (AlO_4^-) anions(65-66). This type of adsorbent will interact with molecules having locally concentrated electron densities, eg., group B and group D molecules.

   3) <u>Specific Adsorbents with Concentrated Electron Densities on Surfaces</u> Example of this type adsorbent would be graphitized carbon black with dense monolayers of group B molecules or macromolecules (polyethylene glycol) deposited on the surface. They could also be adsorbents containing a functional group, such as a cyano, nitrile or carbonyl group on the surface.

   These interactions of sorbate molecules and sorbent types are tabulated in Table I for quick referral.

   In GSC, it is convenient to use n-alkanes as standards because they can easily orient themselves along the sorbent surface. The n-alkanes belong to group A (non-specific molecular interactions) and can be used to compare the capacity of different sorbents regarding non-specific sorption.

   By modifying or changing the solid surface chemistry, one can produce an assortment of both specific and non-specific sorbents for use in GSC. This results in column selectivity at high effectiveness and proper chemical and thermal stability.

   Kiselev expanded his classification of the sorbents on the basis of structural types(67) in addition to his classification on basis of interactions. These classifications are shown in Table II.

   The type I sorbents come with a wide range of specific surface areas. They may vary from 1 to several hundreds of m^2/g. Compressed tablets, made from primary particles of non-porous particles, usually are wide-pore sorbents; the pore size depending upon the pressure under which the tablets have been formed(68).

TABLE I

SORBATE–SORBENT INTERACTIONS

SORBATE GROUP	SORBENT TYPE		
	1. No Surface Active Groups or Ions	2. Locally Concentrated Positive Surface Charges	3. Locally Concentrated Negative Surface Charges
A. Sigma Bonds or Spherically Symmetric Shells	N.S.	N.S.	N.S.
B. Pi Bonds or Electron Density Locally Concentrated on Periphery of Bonds	N.S.	N.S. + S.	N.S. + S.
C. Locally Concentrated Positive Charges on the Periphery of Links	N.S.	N.S. + S.	N.S. + S.
D. Functional Groups Having Both Electron Density and a Positive Charge Concentrated on the Peripheries of Adjacent Links	N.S.	N.S. + S.	N.S. + S.

N.S. = Nonspecific Interactions (mainly dispersion forces); S. = Specific Interactions

The commonly marketed wide pore silica gels fit the type II description. For medium-boiling liquids and especially high-boiling liquids one should use sorbents of type II which have reduced surface areas and larger pores. Hydrothermal treatment or hot steam treatment, of silica gels, will cause pore widening (thousands of angstroms) and surface area reduction (25-50 m^2/g or less)(69-71).

The pore openings of a porous crystal are usually equal in size and thus well adapted for separations based on molecular size (molecular-sieve actions). If the sorbate molecules are too large to penetrate the openings, the porous crystals then behave as a non-porous crystal (type I) towards them. Type IV sorbents are not widely used in GSC because of their large number of strongly adsorbing pores. Macroporous polymers (porous mineral adsorbents) obtained by vacuum sublimation of the frozen intermicellar liquid, usually retain the structure of the initial alko-, benzo- and hydro-type gels from which they were made(72). The same type process may also be used to prepare organic polymers having wide pores, eg., polymer aerogels. The problem with these aerogels of organic polymers for GSC is that they swell in an environment of organic vapors.

The chemical nature of the sorbent surface is not the only cause of adsorption-pore structure has a large influence, also. It has been shown, by static adsorption studies (hydrocarbon vapors on silica gels), that adsorption and heats of adsorption increase with decreasing pore size. This increase is more rapid as the carbon atoms per molecule increases(73-74). The GSC separation of C_1-C_4 hydrocarbons became more complete as the specific surface area increased and the pore size decreased for fine pore silica gels(75). This is related to the increased potential of nonspecific interactions (dispersion forces) as the sorbent pores become narrower. The pores of a solid affect the mass transfer of sorbate and thus will cause the peaks to become more diffuse.

Therefore, in GSC, retention of sorbate components is determined by:
1) The geometric pore structure of the sorbent surface and its chemical nature.
2) The geometric and electronic structures and

molecular weights of the sorbate molecules.
    3) The column temperature

As a consequence, the column separating power depends
upon selectivity of the sorbent and diffuseness or spreading of chromatographic bands moving across the sorbent
layer. Other things being equal, a chromatography column
will be most effective when the bands are less diffuse.
Bands diffuseness may be due to thermodynamics, kinetics,
injection and diffusion causes. These may be summarized
as:

    1) Nonsymmetrical band spreading may be attributed
to nonlinearity of the equilibrium adsorption isotherm,
i.e., deviation of isotherm from Henry's Law. This will
cause the sorbate to move through the column with sorbent
at different rates (rate dependent upon sorbate concentration).

    2) Diffuseness can be attributable to the various
diffusion processes occuring during the transport of the
sorbate through the column. The diffusion processes are
very complex(76-78). This complexity being due to: a)
ordinary diffusion in the gas phase, b) band movement
through particle layers differing in size and shape and
packed in different ways. This will cause diffuseness
related to a non-uniform distribution of gas flow rates
over each cross-sectional area, c) difference in local flow
rates from the average flow rate through the column.
Columns of this type will exhibit what is referred to as
"wall effect". This means that flow at the walls is higher
than the average of the column because resistance at the
wall is less. The great effectiveness of capillary columns
is due mainly to the absence of specific diffusion processes caused by particle layers. However, one does observe
diffuseness in capillary or unpacked columns. This is a
result of the parabolic velocity distribution over the
column cross-section. The velocity is higher at the center
and lower near the walls than the average velocity of the
band.

    3) Diffuseness could be a result of slowness of the
sorption-desorption process, i.e., slow mass transfer or
exchange at sorbent surfaces. Diffuseness in this case can
be non-symmetric because the rates of sorption and desorption are not the same. The band spreading due to the
final rate of mass exchange is closely related to the
diffusion phenomena. Physical adsorption, for all practical

TABLE II

SORBENTS CLASSIFIED BY TYPE

TYPE	EXPLANATION
İ-Non-Porous	Non-porous nono- and polycrystalline sorbents, eg., graphitized carbon black, NaCl. Porous amorphous sorbents such as Aerosil and thermal blacks.
II-Uniform Wide Pores	Large pore glasses, wide pore Xerogels and compressed powders made from non-porous particles (>100A in size and specific surface areas <300 m^2/g).
III-Uniform Fine Pores	Amorphous fine pore Xerogels, fine pore glasses, many activated charcoals and porous crystals (type A and X Zeolites).
IV-Non-uniform Pores	Chalk-like silica gels obtained by hydrolyzing salts of strong acid in a silicate solution.

purposes, is instantaneous. However, the overall process of sorption consists of several parts; a) movement of sorbate molecule toward sorbent surface, this results from intergrain diffusion (outer diffusion), b) movement of sorbate molecules to inside of pores, i.e., internal diffusion of the sorbate molecules in the pores and surface diffusion in the pores, c) the sorption process proper.

4) Diffusion may be influenced by the time it takes to inject sample.

So be cautious in the proper selection of the sorbent-one which gives linear isotherms is essential to minimize diffuseness.

The effects of the causes of diffuseness can be estimated only approximately due to heterogeneity of size and shape of the particles of the sorbent, porosity and non-uniformity of the packing, the heterogeneity of the particle surfaces and the accessibility of these sorbent particles. Mass exchange rate at the solid surface depends on the nature and kinetic energy of the sorbate molecules and

the flow characteristics of the column. External mass exchange can be achieved by conventional diffusion in laminar
flow but the carrier gas is subjected to positive mixing
in turbulent flow.

     To experimentally study the effect of various fac-
tors which cause band spreading in a gas-solid column, one
should eliminate or minimize all factors except the one to
be studied. A good starting point would be to use an
empty column. To calculate the height equivalent to a
theoretical plate, H, under these conditions one can use
the following equation(79):

$$H = 2D/U + 1/24 \cdot r^2 U/D \tag{4}$$

D = coefficient of molecular diffusion
U = linear velocity of gas phase
r = radius of tube

One may evaluate the coefficient of molecular diffusion,
D, from a linear plot of HU versus U^2 by a gas chromatographic procedure. The effects of eddy diffusion and
band broadening resulting from nonuniform distribution of
the gas velocity can be determined from a plot of H versus
U of columns of equal diameter packed with nonporous particles and injecting an unadsorable substance into the
carrier gas.

     The indefinite geometry of the gas-solid system makes
it difficult to develop the kinetics of exchange and
dynamics of column operation. One should compare GSC data
with static method adsorption data to gain appreciation
for their GC data. Similar results can be obtained on homogeneous nonporous surfaces. Very good agreement is obtained when the homogeneous, nonporous and nonspecific
sorbent graphitized thermal carbon black is used(80). Heats
of adsorption, obtained by GC can be 15-20% lower than
those obtained by calorimetry because of nonattainment of
thermodynamic equilibrium, adsorption of water, ammonia,
carbon dioxide and organic substances from the air while
packing the column, adsorption of water from carrier gas,
adsorption of carrier gas molecules on packing active
sites, temperature effects (static experiments are usually
performed at lower temperatures than GC) and sorbate-sorbate associations, which are stronger in static exper-

iments at higher coverages and lower temperatures. Most
of these effects can be eliminated or minimized by using
nonspecific sorbents at high temperatures or specific sorbents which are sufficiently homogeneous and possess high
specific surface areas.

ADSORPTION MOLECULAR THEORY AND GAS-SOLID CHROMATOGRAPHY

Adsorption molecular theory includes molecular-statistical calculation of equilibrium constants, virial coefficients, retention volumes at zero, low and average
surface coverage levels and thermodynamic-constants (heat
and entropy of adsorption and heat capacity of sorption
system).

Analytical applications of this theory make the ad-
sorbent-adsorbate interaction insignificant because of
highly sensitive detectors coupled with low concentrations
of adsorbed molecules and fairly high temperatures. Thus,
in GSC (from molecular theory of adsorption) one need only
consider the adsorbate-adsorbent interactions (Henry's constants and their dependence on temperature). When using
higher concentrations of sorbate, one introduces the next
virial coefficient to take into account the adsorbate-
adsorbate interaction in the adsorbent field.

Potential energy of adsorption of a molecule, in its
most favorable orientation on the surface, is equal to the
heat of adsorption at absolute zero. Due to the slight
dependence of heat of adsorption on temperature (ΔH vs T),
the value of potential energy of adsorption is essentially
the heat of adsorption at temperature of column.

The most convenient method for the estimation of potential energy of adsorption (compounds of low volatility)
is the GC determination of retention volumes at different
temperatures. Precision of such determinations is somewhat low so repeated measurements and statistical analyses is necessary.

Examples of adsorbents which have been used for
molecular adsorption theory studies are:

1) Graphitized Thermal Black

2) Boron Nitride
3) Phthalocyanine Crystals
4) Non-porous Ionic Adsorbents
5) Zeolites
6) Pure Silica and Silica Containing Impurities
7) Silica Modified by Grafted Silylalkyl Groups

The relative role of a sorbate and specific interactions of adsorption on molecular crystals and monolayers containing functional groups can be quite large. This is because the contribution of non-specific interaction energy is small.

Calculations of the retention volumes as a function of concentration (adsorption isotherms) or at zero surface coverage (Henry's constants) is of interest.

$$V_g = K_1 RT \qquad\qquad\qquad (5)$$

V_g = retention volume per unit surface area = V_R^1/A

K_1 = Henry's Constant

The problem of molecular-statistical and thermodynamic estimates of the separation of components leaving the column and the order of appearance of the compounds is of great interest in GC. This can be justified on the fact that the order of emergence of the peaks of two compounds depends on the nature and the geometry of their molecules, the temperature and the sorbent surface. Values for $\ln V_g$ for two components, on the same adsorbent, are usually in the same order (numerically) as the heats of adsorption values.

In the case of low surface coverage, frequently one encounters inversion of emergence order, for two components from the column. For this case, Henry's constant may be calculated at the point of inversion of the plots ($\ln V_g$ vs $1/T$).

$$\ln K_1 = \ln(V_g/RT) = Q_1/RT + [\Delta\bar{S}/R + \ln \tau/p^o] \qquad (6)$$

where $V_g = K_1 RT$

$Q_1 = -RT^2 \partial \ln K_1/\partial T$ = isoteric heat of adsorption

$\Delta \bar{S}$ = R[inK_1 + T∂lnK_1/∂T - lnτ/p^{o}] = differential molar change in entropy of sorbate.

τ = Gibbs adsorption per unit surface area of sorbent
p^{o} = standard pressure

The last term, $\Delta \bar{S}$ + Rln(τ/p^{o}), is the entropy of adsorption and is usually negative. This term, in addition to K_1 and Q_1 is independent of τ. However, all values: K_1, V_g, Q_1 and $\Delta \bar{S}$/R + ln(τ/p^{o}) are temperature dependent. An increase in the isoteric heat of adsorption (Q_1) will result in increases in Henry's constant (lnK_1) and absolute value of entropy of adsorption; in fact, there is often a linear relationship between these values and Q_1.

SORBENTS USED IN GAS–SOLID CHROMATOGRAPHIC STUDIES

We will briefly discuss some of the more commonly used solids which have been used in GSC. This section is not intended to be an-all-inclusive coverage; but only to highlight some of the advances in this interesting area of chromatography.

CARBON SUPPORTS

Carbon, in its many different forms, has been one of the primary solids used for studies of adsorption. Many of the initial studies of adsorption, in general, have been carried out on carbon surfaces. Treating of carbon blacks to temperatures of 3000°C or greater converts the small carbon black particles into polyhedra having homogeneous basal graphite faces(81). This is particularly true of thermal blacks with small specific areas(6-30m^2/g). This basal form of carbon is a crystalline structure similar to graphite and possesses a very homogeneous surface.

The graphitized carbon blacks fall into the category of nonspecific adsorbents because the surfaces carry no functional groups or ions and no unsaturated bonds. Thus, most of the interactions of this sorbent are due to dispersion forces; although some paramagnetic sites are found on the surface. Because of these nonspecific interactions, high heats of adsorption results when the absorbate possesses the proper geometry. Heats of absorption measured

by gas chromatographic methods show closer agreement than
those reported by the limiting isoteric heat of adsorption
at zero coverage measured by static methods(82). Heats of
adsorption of "reference molecules" (hydrocarbons), which
have been determined by either gas chromatographic methods
or by static methods are in excellent agreement and close
to the values of adsorption energies derived by theoretical calculations. However, the heats of adsorption of
alcohols are close to the theoretical values of adsorption
energies but 5 Kcal/mole lower than calorimetric heats of
adsorption. This difference can be attributed to the hydrogen bonding between the alcohol molecules.

Kiselev et al(83) have determined the absolute retention volumes, at various temperatures, of molecules of
different geometric and electronic structures on graphitized carbon blacks. Differential heats of adsorption,
equilibrium constants and other physicochemical properties were then calculated from these data.

Halasz and Horvath(84) were the first to use graphitized carbon black in GSC as a thin-porous layer on the
walls of capillary columns. They were able to obtain a
quick and unusual separation of the xylene isomers; m- and
p-xylenes were perfectly resolved while o- and p-xylenes
remain unresolved. This is the opposite of the results
observed in GLC.

DiCorcia and Bruner(85) were able to separate several
aliphatic and aromatic amines (at several temperatures)
on hydrogen treated graphitized carbon black (Sterling,
M.T., 8m^2/g).

Vidal-Madjer et al(86) used highly efficient open
tubular capillary columns with a thin-layer of graphitized
carbon black to separate high boiling compounds. They
were successful in separating many polynuclear hydrocarbons with a selectivity not obtainable in GLC. Compounds with as high as 22 carbon atoms and five rings were
analyzed in 12 minutes at 585°C.

GELS AND OXIDES

Intensive studies are continuing on traditional ad-

sorbents such as silica gels, alumina gels, and ferrogels
as well as the corresponding oxides, eg., quartz and
alumina. Most important property of surface gels and
corresponding oxides is their degree of hydration and chemical nature of these hydrated surfaces. This property essentially determines their sorptive properties to sorbates which can easily form hydrogen bonds.

    Silica Gel. Silica displays specific adsorption of
molecules because of the high concentration of free hydroxyl groups with partly protonized hydrogen on the hy-
droxylated surface. This is especially true in the case
of sorbates with high electron densities. Polar molecules containing atoms of oxygen or nitrogen with lone
pairs of electrons, alcohols, ethers, ketones, ammonia,
amines, pyridine, and non-polar molecules with pi bonds
(aromatic and unsaturated hydrocarbons) which are polar-
izable are easily separated. If the molecules possess
spherical symmetrical electron shells (inert gases) or only
sigma bonds (saturated hydrocarbons), then the sorbate-
sorbent interactions originate from the dispersion forces.
Nonlinearity of the adsorption isotherm can be reduced by
using wide pore silica gel. Silica gels having average
pore diameters of 20A or less should be used for the
separation of low boiling gases; average pore diameters
from 50 to 200A should be used to separate hydrocarbons
not boiling higher than 100°C. Silica gels with larger
pore diameters, >500A, should be used to analyze higher
boiling substances(69-70).

    Wide pore silica gels with homogeneous surfaces of
high purity are necessary for analytical GSC. Unsymmet-
rical peaks and longer retention times result from silica
gels which contain admixtures of $Al_2O_3 \cdot Fe_2O_3$ and other
oxides. Silica gel surfaces are usually covered by large
amounts of water which is bonded to the silicon atoms.
Activating the silica gel surface to 200°C appears to remove most of this type of water leaving a surface with
hydrogen bonds formed between the surface hydroxyl groups.
Activation to 500°C removes most of the coordinated water
molecules and increases hydrogen bonding between the hydroxyl groups. These hydrogen bonded hydroxyl groups appear
to condense at 600°C to form surface siloxane linkages(87).
Feltl and Smolkova(88) measured the adsorption isotherms of

benzene, n-hexane and cyclohexane by frontal chromatography on porous silica beads of varying surface areas.
They concluded that a change in the geometrical surface
area and porosity played a role in the adsorption process and that the concentration of the hydroxyl groups
(on the surface) is independent of the surface area.

An interesting separation on corroded silica layered
in glass capillaries was achieved by Mohnke and Saffert
(89). Using a 80m capillary column they were able to
separate all the isomers of hydrogen at 77.6°K(-195.6°C).
Using neon as the carrier gas (2cm^3/min) the order of
separation was He, p-H_2, o-H_2, HD, o-D_2 and p-D_2. A similar system was used to separate the neon isotopes(90).

Alumina. Activated alumina has been used as an adsorbent in liquid-solid chromatography more than it has
in GSC. The reason being that the resulting chromatographic peaks are very often asymmetrical due to the nonlinearity of the sorbate isotherms(91). Most of the commercial aluminas that have been used in chromatography are
of the low temperature variety; generally, the impure gamma
alumina with surface areas between 100 to 200 m^2/g. The
active sites on the sorbent alumina are 1) the oxide ions,
2) the Lewis acid sites, i.e., the aluminum ions attached
to three oxygen atoms, and 3) the hydroxyl groups, of
which there are several types depending on the number of
nearest oxygen neighbors(92). It is generally agreed that
the Lewis acid sites are the most active sites on the
alumina surface. This has been determined from catalytic
studies. These authors undertook an infrared spectro-
scopic-gas chromatographic study of the alumina surface.
Three compounds were used as sorbates, n-butanol, pyridine
and acetic anhydride. They found that the aluminum atoms
(electron accepting) were the cause of the severe tailing
and excessively long retention times for the sorbates.
They found that these active sites can be removed by the
use of the pyridine or the acetic anhydride; which in turn
resulted in improved resolution of hydrocarbon samples.

Coating an alumina surface with ferric oxide (Fe_2O_3)
resulted in the quantitative separation of the hydrogen
isomers (H_2, HD, HT, D_2, DT and T_2). This separation was
accomplished using a 3 meter x 4 mm i.d. capillary column
packed with the coated alumina sorbent(93).

Zeolites. Many investigators have studied the synthetic zeolites A and X types for their porous structures
and adsorption properties. These sorbents have rigid
aluminosilicate frames, of which the shape, dimensions,
pore interrelation and surface nature are well known.

The shape of the zeolites is nearly spherical; diameter of zeolite A cavities are 11.4A and the type X
zeolites are 11.6A in diameter. Cavities of the small
zeolites are connected to larger cavities by 6-membered
oxygen windows having diameters of 2.5A. Large cavities
of the type A zeolites are connected to each other by
means of 8-membered oxygen windows, with 4.2A diameters.
Type X cavities are joined by 12-membered oxygen windows
of 8-9A diameter. These small and large cavities form
the primery porous structure of dehydrated zeolite crystals. Accessibility of sorbed molecules is controlled
by diameters of the sorbate and effective diameters of
adsorbents.

Molecules of substances with diameters smaller than
those of the windows, are easily adsorbed. As the sorbate
diameter increases the adsorption rate decreases. Dif-
fusive transfer of molecules to the primary porous
structure shows an increasing activation energy as molecular diameter increases. When molecules become too large
in size, the windows are no longer accessible and adsorption takes place on external surfaces of the zeolite
(ca 10 m^2/g). Thus, selectivity of zeolites is due either
to accessibility or unaccessibility of the porous structure (molecular sieve effect) and sharp differences in
adsorption kinetics resulting from activation energy of
intra-diffuse transfer of sorbed substance into primary
pore structure. Barrer(24) presented a discussion of the
molecular sieve effect and practical applications. The
synthetic zeolite crystals of type A and X (particles
about 1 μ) are usually formed into granules, tablets or
pellets (10-15% binder added) and exhibit low adsorption
properties.

Since the inner surface of the cavity of porous crystals of cationated zeolites carry positive charges concentrated on the exchanged cations they are adsorbents
which interact specifically with molecules of high electron
density (class IV-nonuniform pores). Habgood(94) studied

the retention volumes of oxygen, nitrogen, methane, ethane,
propane, butane, ethylene and propene over the temperature
range of 25° to 400°C. He used the Li^+, Na^+, K^+, Mg^{2+},
Ca^{2+}, Ba^{2+}, and Ag^+ ion-exchanged forms of zeolite X. He
concluded that the cationic field increases with decreasing ionic radius; that divalent cations tend to be found
on less exposed sites and that the polarizing power of
the silver ion is very strong.

Some of the applications of zeolites in GSC have been;
1) analysis of hydrogen isotopes, 2) analysis of low boiling gases, and 3) separation of high molecular weight
hydrocarbons.

Inorganic Adsorbents

Many different types of inorganic compounds have been
used in GSC for the analysis of saturated, unsaturated and
aromatic hydrocarbons. Most of these investigations have
employed inorganic coated columns, i.e., a solid support
such as alumina, silica, carbon, etc. Most investigators
have not used the inorganic compounds directly as packings because of their low surface areas. However, we have
been fairly successful in our laboratory utilizing the
inorganic compounds solely as the packing(95-98). We will
refer to these more specifically a little later in the
section.

Two types of interactions usually occur when using
modified solid packings. 1) Interactions due to dispersion forces which are termed non-specific interactions(62).
This type of interaction is seen when the sorbate molecules
possess either spherically symmetrical electron shells or
sigma bonds. 2) Interactions due to sorbate molecules
having isolated sites, individual bonds, or a system of
bonds of high electron density. In this group we would
find molecules with pi electron systems, lone electron
pairs and related functional groups; all would interact
specifically with the sorbent surface(99). One can determine functional group interactions by comparing the behavior
of two molecules which are similar in structure and physical properties; this assumes that they possess the same
non-specific interactions with the sorbate surface. The
difference in the retention volumes of the two compounds

TABLE III

INTERACTION EFFECTS WITH COMPOUNDS OF
SIMILAR BOILING POINTS

	Compound	Boiling point, °C	Dipole moment, Debye	$\underline{V}_R$, cm^3
A.	Ethylbenzene	132	0.6	52.8
	Chlorobenzene	136	1.7	38.8
B.	Cumene	152	0.8	72.1
	Bromobenzene	155	1.7	59.4
C.	o-Chlorotoluene	159	1.9	79.5
	m-Chlorotoluene	161	1.8	72.4
	p-Chlorotoluene	159	1.9	79.5

COLUMN: 10% NaCl coated F-1 Alumina at 250°C

will then be a measure of these specific effects. We have
illustrated the effect of specific interactions by use of
a CoCl$_2$ and MnCl$_2$ columns(97-98). Gas-solid chromatographic data for some aromatic compounds on a modified
alumina column illustrate some interesting facets of surface
interactions(99). These data are shown in Table III.

As can be seen from the table, dipole-dipole interactions are not the deciding factor. If it were, then
chlorobenzene would elute after ethylbenzene and bromo-
benzene would elute after cumene. Since the retention
volumes are in the reverse order one can speculate that it
is the pi electron density in the rings that determines
the order of elution. The halogen groups deactivate the
ring whereas the alkyl groups activate the rings. Also,
molar refraction data indicates that polarizability of the
molecules may be a contributing factor. We have observed
similar effects with aromatic compounds when using CoCl$_2$
and MnCl$_2$ packings(97-98).

In our investigation of the alkali metal nitrates, as
column packings(95), it was found that the order of decreasing retention volume for all compounds studied was,
CsNO$_3$>RbNO$_3$>LiNO$_3$>KNO$_3$ = NaNO$_3$. Overall several conclusions were reached regarding the use of alkali metal

nitrates as packings in GSC columns:

1) Alkali metal nitrates retard elution greater than
the alkali metal chlorides.

2) Retention volume generally followed boiling points;
two compounds with same boiling point will exhibit greater
retention volume for the more polar compound.

3) Chain branching usually resulted in decreased
retention volumes.

4) Electron releasing groups enhanced adsorption
whereas electron withdrawing groups hindered adsorption.

5) Ortho substituents, in aromatic compounds, cause
stronger effects than meta substituents.

Results of studies of the columns prepared from the
halides of barium and strontium revealed(96): a) Sample
size for overloading generally increased with increasing
anion size of salt, b) Retention volumes generally followed boiling point; polar compounds exhibiting higher retention volumes than a corresponding non-polar compound
with the same boiling point, c) Chain branching reduced
retention volumes and d) Adjusted retention volumes increased as the size of the anion increased.

Studies were then extended(97) to the use of
vanadium, manganese and cobalt salts, namely the chlorides.
These chlorides were chosen because of their number of
available "3d" electrons, i.e., $V(II)(3d^3)$, $Mn(II)(3d^5)$
and $Co(II)(3d^7)$. Various saturated and unsaturated organic
compounds were investigated as adsorbates to observe the
influence of varying pi-electron densities. The heats of
sorption were calculated and found to vary directly with
the pi-electron density of the adsorbate and vary inversely with the number of 3d electrons of the adsorbent.
Separations were a result of the pi-electron density of
the samples. Conjugated systems adsorbed on the salts
were studied but the columns appeared to be best utilized
for compounds having isolated pi-bonds. A highly electro-
negative group, not near a pi-bond, appeared to have little
effect on the degree of sorption. Although it was not our
purpose, at the time, to determine absolutely the mechanism
of sorption, our data did indicate that chemisorption
played a major role in the interaction between adsorbent
and adsorbate. Continued investigation into these salts
(98) indicated that an interaction occurs between sorbates
possessing pi-electron density and the salts. This inter-

ction can be used to separate sorbates varying in their
egree of pi-electron distribution. Thermodynamic cal-
ulations such as enthalpy and entropy of adsorption,
raditionally employed to interpret the nature and degree
f interaction, were inadequate. In this study, the
ifficulty arose primarily because differences in magni-
ude of thermodynamic values were not great enough to
istinguish selectively various mechanism of interaction
r even that interaction did, in fact, occur. Instead,
R spectroscopy was used to establish the fact that in-
eraction involving pi-electron density and the transi-
ion metal salt occured in an ambient, static study. These
esults were used to rationalize the behavior of the same
orbate-sorbent interaction in a dynamic gas chromato-
raphic system.

Barium sulfate, $BaSO_4$, shows high specificity as an
dsorbent (Type 2 adsorbent) for aromatic type compounds.
oth barium and magnesium sulfates are good adsorbents for
he separation of B group molecules (undergo weak specific
nteractions with adsorbents of 2nd type(100).

Organic Adsorbents

Organic type adsorbents have also been used in GSC.
hese may be sorted into four categories:

1) Organic crystal compounds. Examples of this type
orbent would be benzophenone on firebrick(101), an-
hraquinone on graphitized carbon black(102), copper py-
idine complexes(103), 1,3,5-tri-nitrobenzene on Chromo-
orb(104), phthalic anhydride and phthalic acid isomers on
hromosorb G(105), and nickel complexes on Chromosorb W
106).
2) Conventional liquid stationary phases below their
elting points. Included in the category one would find
E-30 on Chromosorb(107), Carbowax 20M on Chromosorb(108),
nd solid ureides of N-TFA-1-phenylalanine cyclohexyl
sters(109).
3) Organic clay derivatives. This would include
entone 34 and any other modified organo clays. The cations
n the naturally clays are replaced by alkyl quaternary
mmonium ions and the resulting compounds show a selective
etention of aromatic compounds relative to paraffins and
aphthenes(110-116).

4) Porous polymers. Hollis(117) was the first to
use porous co-polymers of the ethylvinylbenzene-divinyl-
benzene type in GC. Since then many macroporous copolymers
have been used as packings, especially the co-polymer of
styrene and ethylvinylbenzene with divinylbenzene as the
cross-linking agent. These porous polymers are manufactured under various trade names, eg., Porapak, Chromosorb 101 and 102, Par 1 and 2, Synachrom, Polysorb, etc.
These types of packings are very good for traces of polar
compounds, gaseous components in air, amines and
fluorinated compounds. An excellent coverage of these
packings is covered in an article by Vidal-Madjar and
Guiochon(118). In our own laboratory we found these packings to be very good for the separation of solid organo-
phosphorus compounds(119). In this study it was found
that coating the porous polymer beads with SE-30 resulted
in a better separation of components, as well as sharper
peaks. Heats of adsorption calculations showed that the
separations were more due to partitioning rather than due
to adsorption.

THERMODYNAMICS OF ADSORPTION
AND ITS RELATION TO GAS-SOLID CHROMATOGRAPHY

A convenient way to present the processes taking place
at the gas-solid interface, is by the use of the thermodynamics of adsorption. In this manner we have a better
way of expressing what may account for the interactions
at the surface. This is true when we attempt to interpret
gas-solid chromatographic data. A term much used in
chromatography is the retention volume, i.e., the volume
of carrier needed to transport the sample molecule through
the column and into the detector system. In GSC, we refer
to the retention volume in the same way but the sample
molecule is more specifically referred to as the sorbate.
Nomenclature between GSC and GLC does not necessarily
agree. In GLC, retention volume is represented by V_R
(meaning the total retention volume). V_R is used in GSC
to represent corrected retention volume(120; whereas corrected retention volume in GLC is given the symbol V_R^0.
Specific retention volume in GSC is V_s^T, but in GLC it is
V_g. They are defined differently:

1) In GLC, specific retention volume, V_g is:

$$V_g = V_N \; 273/w_L \, T \tag{7}$$

V_N = net retention volume
w_L = mass liquid phase $(V_L \rho_L)$
T = absolute temperature in $^\circ$K

2) In GSC specific retention volume, V_s^T is (120):

$$V_s^T = V_R/A \tag{8}$$

A = surface area of sorbent

Thus, in our discussion we will refer to corrected retention volume as V_R^O and specific retention volume as V_g. Our reason for staying with the GLC symbols is two-fold: GLC nomenclature is more widely known and recognized and GSC data has not been put on a firm basis as yet and there are still some ambiguities.

The distribution constant for the sorbate in equilibrium at the solid surface and in the vapor phase may be defined as:

$$K_D = C_s/C_g = \text{moles/m}^2 \div \text{moles/cm}^3 = \text{cm}^3/\text{m}^2 \tag{9}$$

Equation (9) would also define the corrected retention volume, V_R^O, with respect to the sorbent surface area and the specific retention volume, V_g:

$$K_D = C_s/C_g = V_R^O/A = V_g \tag{10}$$

If one expresses C_s in terms of moles/cm^2 then the distribution constant may be redifined as, K' (1m^2 = 10^{4}cm^2):

$$K_D' = \text{moles/10}^4\text{cm}^2 \div \text{moles/cm}^3 = \text{cm}^3/10^4 \text{ cm}^2$$
$$= 10^{-4} \, K_D(\text{cm}^3/\text{cm}^2) \tag{11}$$

The free energy of adsorption, $-\Delta G_{ads}'$, is related to the distribution constant by

$$-\Delta G_{ads}' = RT\ln K_D = RT\ln V_g \tag{12}$$

which in turn is

$$-\Delta G_{ads}' = -\Delta H_{ads}' + T\Delta S_{ads}' \tag{13}$$

Equations 12 and 13 may be rearranged to give specific retention volume, V_g

$$\log V_g = -\Delta G'_{ads}/2.3RT = -\Delta H'_{ads}/2.3RT + \Delta S'_{ads}/2.3R \quad (14)$$

as a function of enthalpy and entropy of adsorption.

If one then assumes idealized standard states, it is
possible to determine the standard free energy (ΔG^o_{ads}) and
the standard state entropy (ΔS^o_{ads}) from GC retention data.

1) The standard state of the sorbate, in the gas
phase, may be defined as a partial pressure of 1 atmosphere
with sorbate vapors behaving as a perfect gas. Thus, we
can use the equation, $PV = nRT$.

2) The standard state of the sorbate at the surface
(in the sorbed state) can be assumed to be a two-dimension-
al perfect gas at 1 atmosphere(121). From this it follows
that the mean distance between sorbed molecules is defined
to be the same as in the three-dimensional gas phase standard state. Solving for the intermolecular distance, one
can evaluate the area per molecule at standard conditions.
In order to do this we must calculate a conditional gas
constant, R', which is consistent with our assumptions.
Thus

$$PA = nR'T \quad (15)$$

A = surface area in lieu of V for volume
R' = conditional gas constant for these conditions

$$P/R'T = n/A \quad (16)$$

The surface space occupied by 1 molecule would be

$$
\begin{aligned}
(RT/N)^{1/3} &= (82.05 \times 273.16/6.023 \times 10^{23})^{1/3} \\
&= (37.212 \times 10^{-21})^{1/3} \\
&= 3.34 \times 10^{-7} \text{ cm} \quad (17)
\end{aligned}
$$

The area per molecule may then be calculated as

$$(3.34 \times 10^{-7} \text{cm})^2 = 11.16 \times 10^{-14} \text{cm}^2 \quad (18)$$

and the area per mole as

$$\text{Area per molecule} \times N = (11.16 \times 10^{-14})(6.023 \times 10^{23})$$
$$= 6.719 \times 10^{10} \text{ cm}^2 = A \quad (19)$$

Solving equation 15 for R' gives

$$R' = PA/nT = (1 \text{ atm})(6.71 \times 10^{10})/(1 \text{ mole})(273.16)$$
$$= 2.460 \times 10^8 \text{ cm}^2 \text{ atm/mole} \quad (20)$$

Insert value of R' into equation 16

$$P/R'T = n/A = 1 \text{ atm}/(2.46 \times 10^8)T$$
$$n/A = 4.07 \times 10^{-9}/T \text{ (moles/cm}^2) \quad (21)$$

We are then led to a standard state surface concentration, C_s^o, which is defined by

$$n/A = 4.07 \times 10^{-9}/T = C_s^o \quad (22)$$

where T is the column temperature.

The combination of equations 9, 11 and 22 will give the gas phase sorbate concentration, C_g

$$C_g = 4.07 \times 10^{-9}/TK_D' \quad (23)$$

Combining equation 23 with perfect gas equation PV = nRT, we obtain

$$P = P_{(equil)} \qquad n = C_g$$
$$P_{(equil)} = C_g RT = \frac{4.07 \times 10^{-9}}{TK_D'} \cdot RT$$
$$= 4.07 \times 10^{-9} R/K_D' \quad (24)$$

$P_{(equil)}$ = equilibrium partial pressure of sorbate vapor in equilibrium with sorbent.

Thus, the differential molar free energy, ΔG_{ads}^o (energy needed to transfer one mole of vapor at one atmosphere to its equilibrium vapor pressure, $P_{(equil)}$), is

$$\Delta G_{ads}^o = RT \ln(P_{(equil)}/1) \quad (25)$$

substituting equation 24 into equation 25, we obtain

$$\Delta G_{ads}^o = RT \ln(4.07 \times 10^{-9} R/K_D') \quad (26)$$

and substituting equations 10 and 11 into equation 26 gives

$$\Delta G^o_{ads} = RT \ln(4.07 \times 10^{-9} R/10^{-4}K_D)$$
$$= RT \ln(4.07 \times 10^{-5} R/K_D)$$
$$= RT \ln(4.07 \times 10^{-5} R/V_g) \tag{27}$$

$R = 0.08205$ liter atms = 82.05 cm^3 atm

$$\Delta G^o_{ads} = RT \ln((4.07 \times 10^{-5})\, 82.05) - RT \ln V_g \tag{28}$$

from equation 12

$$-\Delta G'_{ads} = RT \ln V_g$$

therefore

$$\Delta G^o_{ads} = RT \ln(4.07 \times 10^{-5} \times 82.05) + \Delta G'_{ads} \tag{29}$$

$R = 1.987$ calories/degree mole, thus

$$\Delta G^o_{ads} = 1.987T \ln(333.9435 \times 10^{-5}) + \Delta G'_{ads}$$
$$= 1.99T(-5.7020) + \Delta G'_{ads}$$
$$= \Delta G'_{ads} - 11.33T \tag{30}$$

From equations 12 and 30

$$\Delta G^o_{ads} = -RT \ln V_g - 11.33T = -2.3RT \log V_g - 11.33T$$
$$\Delta G^o_{ads}/2.3RT = -\log V_g - 11.33T/2.3RT = -\log V_g - 11.33/2.3R$$

$R = 1.99$ calories/degree mole

$$\Delta G^o_{ads}/4.58T = -\log V_g - 11.33/4.58 = -\log V_g - 2.48$$

thus, $$\log V_g = -\Delta G^o_{ads}/4.58T - 2.48 \tag{31}$$

From equations 30 and 31

$$\log V_g = (-\Delta G'_{ads} + 11.33T)/4.58T - 2.48$$
$$= -\Delta G'_{ads}/4.58T + 11.33T/4.58T - 2.48$$
$$= -\Delta G'_{ads}/4.58T + 2.48 - 2.48$$
$$= -\Delta G'_{ads}/4.58T \tag{32}$$

V_g is in cm^3/m^2 and $\Delta G'_{ads}$ is in calories

The log V_g term can be divided into additive components, all of which contribute to the overall adsorption or retention of a component. We can let $(logV_g)_0$ be the zero point value, $(logV_g)_c$ be the contribution per carbon atom (more specifically per methylene group, $-CH_2-$), $(logV_g)_\pi$ the contribution of the pi bond, and $(logV_g)_Q$ the contribution of the substituent on the benzene ring. $\underline{n}$ is the number of carbon atoms and $\underline{m}$ the number of pi bonds. Thus,

$$log\ V_g = (logV_g)_0 + n(logV_g)_c + m(logV_g)_\pi + (logV_g)_Q\ (33)$$

By plotting $logV_g$ versus $\underline{n}$ for normal alkanes one obtains

   a) $(logV_g)_c$ from the slope
   b) $(logV_g)_0$ from the intercept
   c) $(10gV_g)_\pi$ from the difference in $logV_g$ units of an alkene and an alkane with the same number of carbon atoms.
   d) $(logV_g)_Q$ from the difference in $logV_g$ units of benzene and the corresponding substituted benzene molecule.

By determining these parameters of adsorption at three different temperatures, one can calculate the free energy, enthalpy and entropy of adsorption by use of equations 12, 13, 14 (plotting $logV_g$ versus $1/T$ at three temperatures); slope gives $\dfrac{-\Delta H_{ads}}{2.3R}$, ΔG is obtained from equation 12, and ΔS is obtained from equation 13.

CONCLUSIONS

The main developments in GSC have been to solve the most important limitations of GSC, i.e., the lack of adsorbents which could be used for the separation of complex organic molecular mixtures. The chemist now has better control of the purity and homogeneity of the conventional adsorbents, synthetic macroporous adsorbents, surface modified adsorbents from the grafting of various molecules and porous polymer adsorbents. These permit the adjustment of relative contributions of specific and nonspecific forces to the adsorption process, fast and easy separation of closely related isomers; including optical isomers.

The main areas where GSC has been utilized have been:
1) The analyses of gases and low-boiling hydrocarbon mixtures. Adsorbents such as zeolites, silica gels, alumina and porous polymers being the most pertinent.
2) The separation and determination of compounds which cannot be handled easily by GLC. GLC is less specific as compared to GSC. Separation of geometrical isomers would be a good example. Adsorbents utilized have been graphitized carbon black, modified silica gels, as well as various inorganic and organic adsorbents.
3) The separation of very polar compounds which may interact with the supports used in GLC, eg., porous polymers, graphitized carbon black.

GSC has advanced to its present state because of the availability of so many studies into the adsorption process which have provided an understanding of adsorption phenomenon. Retention volumes, in favorable cases, can be predicted by utilizing data obtained from collisions in the gas phase and methods of statistical thermodynamics.

GSC should develop along two pathways; it should become a very valuable analytical, as well as, preparative method for the analysis of gases and mixtures at high temperatures and an indispensable, sensitive and rapid method for investigating solid surfaces and molecular interactions.

In combination with either gas-liquid or gas-solid elution chromatography, gas-solid displacement chromatography provides an effective technique for trace analysis. Gas-solid displacement chromatography may end up as a very useful technique for preparative work. Another outstanding feature of GSC is its great thermal stability as compared to GLC. Solid surfaces or surfaces covered only with a monolayer of modifying medium, will lead to faster and better separations than with liquid phases. In some cases, we can even reverse the order of elution; causing an impurity to elute before a major component. Although solids are more temperature stable than liquid phases, there is an upper limit of 450°-500°C. This limit is determined by the sorbate, on the one hand, and the sorbent on the other. In case of the sorbates, their stability imposes the limit whereas with sorbents their catalytic activity is the limiting factor. In spite of the higher temperatures permissible by sorbent, the technique of GSC does not easily lend itself to the separation of very high molecular weight substances. In this case it might be better to utilize liquid-liquid or liquid-solid chromatography. One could still separate high molecular weight substances if volatile derivatives can be prepared. The preparation of a solid sorbent packing is not as easy as preparing a liquid substrate but for repetitive analyses it should be stressed. A column for GSC can be operated isothermally or temperature programmed for long periods of time without the problems of detector drift or signal fluctuation and stability of the column packing.

Separation of cis- or trans-isomers can be accommodated more readily by GSC and usually requires a smaller number of plates. This type separation results from the nearness of approach, to the surface of the column packing, of the cis-isomer compared to the trans-isomer. This phenomena is readily apparent from the heats of adsorption of representative C_5- C_8 alkanes and alkenes(122). A gas-solid column of graphitized carbon black with an efficiency of 700 plates can separate cis- and trans-isomers of alkylcyclohexanes. The separation of benzene and deutero-benzene is more efficiently performed in a capillary gas-solid column(123,124) than in a capillary gas-liquid column.

In many cases, symmetrical elutions have been obtained for polar compounds. This may be rationalized on the basis that the high sensitivity of present detectors allow sufficiently small samples to be used. This decreased sample size limits the sorbate-sorbate interactions; especially for those molecules capable of intra-hydrogen bonding. The common belief has been that only molecules with a permanent dipole, lone pair of electrons or pi electrons could interact with polar surfaces. Some separations may be based upon utilizing the same properties of the sorbates but interacting with non-polar surfaces. In this manner one would only have to change the strength of dispersion interactions.

The thermodynamic aspects of interactions at the gas-solid interface must and are being studied in relation to column performance. Fortunately, it is very easy to measure ΔH, ΔG and ΔS by GSC.

The high selectivity of physicochemical processes in
GSC depends upon the interaction of sorbate molecules with
the column packing surface. This interaction should be a
weak molecular one and not strong in chemical nature, so
as to preserve the individuality of the sorbate molecules
and permit them to elute in a short period of time. The
surface of column packings is usually very heterogeneous
and difficult to control. A main cause of heterogeneity
is the variation in pore sizes. Elimination of geometric
heterogeneity increases adsorption effectiveness and
selectivity of the column. GSC lends itself to fast
analyses because of the rapid mass exchange at the surfaces.

The great interest shown and the future expansion in
GSC is due to use of inorganic salt adsorbents (95-97),
zeolites(125), modified oxides(101,126) and surfaces of
dense monomolecular polymer layers adsorbed on homogeneous
surfaces of non-porous and wide pore adsorbents(63).

Gas-solid chromatography has become a very sophisticated technique in separations and also in the area of
analytical chemistry. This has been due to the preparation
of solid sorbents with the required porosity, pore distribution, surface chemical composition and specific
surface area. At this point in time the possibilities of
GSC are well beyond those of GLC.

The use of porous adsorbent layers on different supports and as chemically modified adsorbents is going to
become more and more important in the near future. The
various methods of blocking the active centers or changing
the chemical nature of the surface by reactions or by
deposition of a monomolecular layer, anchored or not on
the surface, are developing fast and becoming more sophisticated and flexible.

ACKNOWLEDGEMENTS

The author would like to thank all those people who
made this paper possible. A special thanks goes to the
students who have worked with me in this area of chromato-

graphy over the years. Among that group I would like to
mention George Weinert, Robert Gondek, Thomas Scales,
Eugene McGonigle, Mathew O'Brien, Mary Kaiser, Edward
Smith, Gary McCrae and Joseph Giannovario.

LITERATURE CITED

1) Berl, E., Andress, K., and Muller, W., Z. Angew.
   Chem., 34, 125-127, 278-279 (1921).
2) Berl, E. and Schwebel, W., Z. Angew. Chem., 36,
   541-545, 552-554 (1923).
3) Berl, E. and Wachendorff, E., Z. Angew. Chem., 37,
   205-206 (1924).
4) Peters, K. and Weil, K., Z. physik. Chem., Leipzig,
   A148, 1 (1930).
5) Peters, K. and Lohmar, W., Angew. Chem., 50, 40
   (1937).
6) Peters, K., Z. physik. Chem., Leipzig, A180, 44 (1937)
7) Euchen, A. and Knick, H., Brennstoff-Chem., 17, 241
   (1936).
8) Henjes, R., Oel u.Kohle ver. Erdoel u. Teer, 14,
   1075-1085 (1938).
9) Ferber, E. and Luther, H., Angew. Chem., 53, 31 (1940).
10) Turner, W.C., Oil Gas J., 41, 48, 29th April, 1943;
    Petroleum Refiner 22, 140, May 1943.
11) Bulletin No. 205, Burrell Technical Supply Co.,
    Pittsburgh, Pa. 15200.
12) Claesson, S., Arkiv. Kemi, Min. Geol. A, 23, (1946).
13) Claesson, S., Discussions Faraday Soc., 7, 34 (1949).
14) Cremer, E. and Prior, F., Z. Elektrochem., 55, 66
    (1951).
15) Cremer, E. and Mueller, R., Z. Elektrochem., 55, 217
    (1951).
16) Janak, J., Chem. Listy, 47, 464, 817, 1184 (1953).
17) Janak, J., Collection Czechoslov. Chem. Commun., 18,
    798 (1953); 19, 684 (1953).
18) Patton, H.W., Lewis, J. S. and Kaye, W. I., Anal.
    Chem., 27, 170-174 (1955).
19) Turkeltaub, N. M., Zhur. Anal. Khim., 5, 200 (1950).
20) Cremer, E. and Roselius, L., Angew. Chem., 70, 42
    (1958).
21) Giddings, J. C., Anal. Chem., 35, 1999 (1963).
22) Giddings, J. C., Anal. Chem., 36, 1170 (1964).

23) Grubner, O., in "Advances in Chromatography," Vol. 6, p. 173, J. C. Giddings and R. A. Keller, eds., Marcel Dekker, Inc., New York, 1968.

24) Barrer, R. M., Proc. Royal Soc. (London), A161, 476 (1937).

25) Orr, W. J. G., Proc. Royal Soc. (London), A173, 349 (1939).

26) Avgul, N. N., Berezin, G. I., Kiselev, A. V. and Lygina, I. A., Izv. Akad. Nauk SSSR, Otd. Khim. Nauk, 1304 (1956).

27) Avgul, N. N., Berezin, G. I., Kiselev, A. V. and Lygina, I. A., Zh. Fiz. Khim., 30, 2106 (1956).

28) Avgul, N. N. and Kiselev, A. V., Izv. Akad. Nauk SSSR, Otd. Khim. Nauk, 230 (1957).

29) Avgul, N. N., Isirikyan, A. A., Kiselev, A. V., Lygina, I. A. and Poshkus, D. P., Izv. Akad. Nauk SSSR, Otd. Khim. Nauk, 1314 (1957).

30) Kiselev, A. V. and Khropova, E. V., Izv. Akad. Nauk SSSR, Otd. Khim. Nauk, 389 (1958).

31) Avgul, N. N., Kiselev, A. V., Lygina, I. A. and Poshkus, D. P., Izv. Akad. Nauk SSSR, Otd. Khim. Nauk, 1196 (1959).

32) Kiselev, A. V. and Poshkus, D. P., Zh. Fiz. Khim., 32, 2824 (1958).

33) Poshkus, D. P. and Kiselev, A. V., Zh. Fiz. Khim., 34, 2640 (1960).

34) Poshkus, D. P. and Kiselev, A. V., Zh. Fiz. Khim., 34, 2646 (1960).

35) Steele, W. A. and Halsey, C. D. Jr., J. Chem. Phys., 22, 979 (1954).

36) Steele, W. A. and Halsey, C. D. Jr., J. Phys. Chem., 59, 57 (1955).

37) Freeman, M. P. and Halsey, C. D. Jr., J. Phys. Chem., 59, 181 (1955).

38) Kwan, T., Freeman, M. P. and Halsey, C. D. Jr., J. Phys. Chem., 59, 600 (1955).

39) Greene, S. A. and Pust, H., J. Phys. Chem., 62, 55 (1958).

40) Kington, G. L., in "The Structure and Properties of Porous Materials," 10th Symposium of the Colston Res. Soc., p. 59, Butterworths, London, 1958.

41) Kington, G. L. and MacLeod, A. C., Trans. Faraday Soc., 55, 1799 (1959).

42) Holmes, J. M. and Beebe, R. A., Can. J. Chem., 35, 1542 (1957).

43) Ross, S. and Pultz, W. W., J. Colloid Sci., 13, 397 (1958).

44) Kiselev, A. V., Dokl. Akad. Nauk. SSSR, 117, 1023 (1957).

45) Kiselev. A. V. Kolloidn. Zh., 20, 338 (1958).

46) Barrer, R. M. and Wasilewski, S., Trans. Faraday Soc., 57, 1140 (1961).

47) Roth, J. F. and Ellwood, R. J., Anal. Chem., 31, 1738 (1959).

48) Roginski, S. Z., Yanovskii, M. I., Peichzhan, L., Gasiev, G. A. Zhabrova, G. M., Kadenatsi, B. M. Brazhnikov, V. V. Neimark, I. E. and Piontkovskaya, M. A., Kinetika i Kataliz, 1, 287 (1960).

49) Miyazaki, K., Seiyama, T. and Sakai, W., J. Chem. Soc. Japan, Ind. Chem. Sect., 59, 146 (1956).

50) Dibbs, H. P., J. Appl. Chem., 10, 372 (1960).

51) Puri, B. R., Mittal, S. and Sharma, L. R., Res. Bull. Panjab Univ., 111, 309 (1957).

52) Schay, G., Nagy, L. G. and Szekreniesy, T. Magy. Kem. Folyvoirant, 66, 271 (1960).

53) Mathews, D. H., J. Appl. Chem., 7, 610 (1957).

54) Hughes, T. H., J. Appl. Chem., 9, 360 (1959).

55) Grubner, O., Collection Czech. Chem. Commun., 25, 180 (1960).

56) Durif, S., J. Chim. Phys., 54, 633 (1957).

57) Alexanian, C., Durif, S. and Soule, J. L., J. Phys. Radium, 20(12), 139 (1959).

58) Freeman, M. P., J. Phys. Chem., 64, 32 (1960).

59) Everett, D. H., in "The Structure and Properties of Porous Materials," 10th Symposium of the Colston Res. Soc., p. 95, Butterworths, London, 1958.

60) Crawford, V., Quart. Rev. (London), 14, 378 (1960).

61) Sheppard, N., Mathieu, M. V. and Yates, D. J. C., Z. Electrochem., 64, 734 (1960).

62) Kiselev, A. V., in "Gas Chromatography-1964," A. Goldup, ed., p. 238, Butterworths, London, 1965.

63) Kiselev, A. V., Disc. Faraday Soc., 40, 205 (1965).

64) Kiselev, A. V., Rev. Gen. Caoutchouc, 41, 377 (1964).

65) Barrer, R. M., in "Structure and Properties of Porous Materials," 10th Symposium of the Colston Res. Soc., p. 6, Butterworths, London, 1958.

66) Barrer, R. M., in "Non-Stoichiometric Compounds," Mandelcorn, L., ed., Academic Press, New York, 1964, p. 309.

67) Kiselev, A. V., "Methods of Investigating the Struc-
     ture of Highly Dispersed and Porous Bodies", USSR
     Academy of Science, Moscow, p. 47, 1958.
68) Venable, R. and Wade, W. H., J. Phys. Chem., 69,
     1395 (1965).
69) Kiselev, A. V., in "Gas Chromatography-1962", M.
     van Swaay, ed., P. XXXIV, Butterworths, London, 1963.
70) Kiselev, A. V., Nikitin, Y. S. Petrova, R. S.
     Shcherbakova, K. D. and Yashin, Y. I., Anal. Chem.,
     36, 1526 (1964).
71) Ries, H. E., Advan. Catalysis, 4, 87 (1952).
72) Vinogradov, G. V., Titkova, L. V. Akshinskaya, N.V.,
     Bebris, N. K., Kiselev, A. V. and Nikitin, Y. S.,
     Zh. Fiz. Khim., 40, 881 (1966).
73) Kiselev, A. V., in the "Structure and Properties of
     Porous Materials," 10th Symposium of the Colston Res.
     Soc., p. 195, Butterworths, London, 1958.
74) Kiselev, A. V., Proc. Intern. Congr. Surface
     Activity, 2nd., Vol. 2, p. 179, London, 1957.
75) Vyakhirev, D. A., Chernyayev, N. P. and Bruk, A. I.,
     Zh. Fiz. Khim., 34, 1096 (1960).
76) "Gas Chromatography," A.I.M. Keulemans, Ed., 2nd Ed.,
     Reinhold, New York, 1957.
77) "Theoretische Grundlagen der Gaschromatographie",
     G. Schay, Akademie-Verlag, Berlin, 1961.
78) Guiochon, G., Bull. Soc. Chim. France, 3367 (1965).
79) Giddings, J. C. and Spencer, S. L., J. Chem. Phys.,
     33, 1579 (1960).
80) Kiselev, A. V., in "Advances in Chromatography,"
     J. C. Giddings and R. A. Keller, eds., Vol. 4, p. 189,
     Marcel Dekker, Inc., New York, 1967.
81) Avgul, N. N. and Kiselev, A. V., in "Chemistry and
     Physics of Carbon," P. L. Walker, ed., p. 1, Marcel
     Dekker, Inc., New York, 1970.
82) Ross, S., Saelens, J. K. and Olivier, J. P., J. Phys.
     Chem., 66, 696 (1962).
83) Kiselev, A. V. and Yashin, Y. I., in "La Chromato-
     graphie Gaz-Solide, p. 26, Masson, Paris, 1969.
84) Halasz, I. and Horvath, C., Nature, 197, 71 (1963).
85) DiCorcia, A. and Bruner, F., Anal. Chem., 43, 1634
     (1971).
86) Vidal-Madjar, C., Ganasia, J. and Guiochon, G., in
     "Gas Chromatography-1970", R. Stock, ed. p. 20, The
     Institute of Petroleum, London, 1971.

87) Cadogan, D. F. and Sawyer, D. T., Anal. Chem., 42, 190 (1970).
88) Feltl, L. and Smolkova, E., J. Chromatog., 65, 249 (1972).
89) Mohnke, M. and Saffert, W., in "Gas Chromatography 1962," M. van Swaay, ed., p. 216, Butterworths, London, 1962.
90) Purer, A., Kaplan, R. L. and Smith, D. R., in "Advances in Chromatography 1969", A. Zlatkis, ed., p. 57, Preston Technical Abstracts, Evanston, Ill., 1969.
91) Snyder, L. R., in "Principles of Adsorption Chromatography," p. 163, Marcel Dekker, Inc., New York, 1968.
92) Neumann, M. G. and Hertl, W., J. Chromatog., 65, 467 (1972).
93) Genty, C. and Schott, R., Anal. Chem., 42, 7 (1970).
94) Habgood, W. W., Can. J. Chem., 42, 2340 (1964).
95) Grob, R. L., Weinert, G. W. and Drelich, J. W., J. Chromatog., 30, 305–324 (1967).
96) Grob, R. L., Gondek, R. K. and Scales, T. A., J. Chromatog., 53, 477–486 (1970).
97) Grob, R. L. and McGonigle, E. J., J. Chromatog., 59, 13–20 (1971).
98) McGonigle, E. J. and Grob, R. L., J. Chromatog., 101, 39–50 (1974).
99) Brookman, D. J. and Sawyer, D. T., Anal. Chem., 40, 106 (1968).
00) Belyakova, L. D., Kiselev, A. V. and Soloyan, G. A., Chromatographia, 3, 254–259 (1970).
01) Scott, C. G., in "Gas Chromatography 1962," M. van Swaay, ed., p. 36, Butterworths, London, 1962.
02) Vidal-Madjar, C. and Guiochon, G., Separation Science, 2, 155 (1967).
03) Rogers, L. B. and Altenau, A. G., Anal. Chem., 35, 915 (1963).; ibid., 36, 1726 (1964) ibid., 37, 1432 (1965).
04) Cvetanovic, R. J., Duncan, F. J. and Falconer, W.E., Can. J. Chem., 42, 2410 (1964).
05) Heveran, J. E. and Rogers, L. B., J. Chromatog., 25, 213 (1966).
06) Pflaum, R. T. and Cook, L. E., J. Chromatog., 50, 120 (1970).
07) Altenau, A. G., Kramer, R. E., McAdoo, D. J., and Merritt, C., J. Gas Chromatog., 4, 96 (1966).

108) Dal Nogare, S., Anal. Chem., 37, 1450 (1965).

109) Corbin, J. A. and Rogers, L. B., Anal. Chem., 42, 1786 (1970).

110) White, D., Nature, 179, 1075 (1957).

111) Mortimer, J. V. and Gent, P. L., Nature, 197, 789 (1963).

112) Spencer, S., Anal. Chem., 35, 592 (1963).

113) Mortimer, J. V. and Gent, P. L., Anal. Chem., 36, 754 (1964).

114) Fuchs, P. J., J. Chromatog., 65, 219 (1972).

115) Taramasso, M. and Veniale, F., Chromatographia, 2, 239 (1969).

116) Taramasso, M., J. Chromatog., 58, 31 (1971).

117) Hollis, O. L., Anal. Chem., 38, 309 (1966).

118) Vidal-Madjar, C. and Guiochon, G., Sepn. and Purif. Methods, 2(1), 1 (1973).

119) Grob, R. L. and McCrea, G. L., Anal. Letters, 1 (2), 53-59 (1967).

120) Sawyer, D. T. and Brookman, D. J. Anal. Chem., 40, 1847 (1968).

121) deBoer, J. H. and Kryer, S., in Proc. Acad. Sci. Amsterdam, 55b, 451 (1952).

122) Scott, C. G., J. Gas Chromatog., 4, 4-7 (1966).

123) Bruner, F. and Cartoni, G. P., J. Chromatog., 10, 396 (1963).

124) Liberti, A., Cartoni, G. P. and Bruner, F., J. Chromatog., 12, 8 (1963).

125) Kiselev, A. V., Chernenkova, Y. L. and Yashin, Y. I., Neftekhimiya, 5, 141 (1965).

126) Bruk, A. I., Vjakhirev, D. A., Kiselev, A. V., Nikitin, Y. S. and Olefirenko, N. M. Neftekhimiya, 7, 145 (1967).

THE USE OF GC-MS TECHNIQUES FOR THE ANALYSIS OF THERAPEUTIC AGENTS IN BLOOD

Michael Lehrer and Arthur Karmen

Department of Laboratory Medicine
Albert Einstein College of Medicine of
Yeshiva University
1300 Morris Park Avenue, Bronx, New York

Combined Gas Chromatography-Mass Spectrometry has become an established method for detecting and quantifying compounds in complex mixtures (1, 2, 3). In the work described here, we studied various methods of applying this powerful technique to the quantitative assay of the anti-epileptic drug, diphenylhydantoin, in serum samples submitted to our clinical laboratory. Mass fragmentography, in which the elution of compounds from the GC column is monitored by monitoring for preselected specific ions or spectra of ions as a function of time was used with both electron impact (EI) and chemical ionization (CI) in order to assess the sensitivity and potential advantages of each technique.

MATERIALS AND METHODS

1. Gas chromatograph: Shimadzu GC-3BF with hydrogen flame ionization detector (American Instrument Co., Inc., Silver Springs, Maryland).
2. GC-MC: Finnigan Model 3200 GC-MS with glass-jet helium separator (Finnigan Corp., Sunnyvale, California).
3. The electron multiplier signal of the GC-MS was recorded as a function of time with a Leeds

Northrup Speedomax XL 680 strip chart recorder
(Leeds & Northrup Co., North Wales, Penn.).

Reagents

1. Trimethylanilium Hydroxide (TMAH) 0.2 M in me-
    thanol (Pierce Chemical Co., Rockford, Illinois).
2. 5-(p-Methylphenyl)-5-phenylhydantoin (MPPH) and
    5,5-diphenylhydantoin (DPH) (Aldrich Chemical
    Co., Milwaukee, Wisconsin).

Stock Standards

DPH and MPPH solutions were prepared to contain
200 ug/ml of methanol.

Calibration Standards

These solutions were freshly prepared immediately prior to use.

1. MPPH (0.4 ug/ul): The residue (400 ug), after
evaporation of 2 ml of MPPH stock standard, was dissolved in trimethylanilium hydroxide (230 ul) and
diluted with methanol (770 ul).
2. DPH (0.5 ug/ul): The residue (200 ug), after
evaporation of 1 ml of DPH stock standard, was dissolved in 0.4 ug/ul MPPH solution (400 ul).

A standard curve was prepared by injecting 1 ul
samples containing 0.5, 0.4, 0.3, 0.2, 0.1, and
0.05 ug of DPH and 0.4 ug of MPPH prepared by dilu-
ting the DPH calibrating standard with MPPH calibrating standard.

Procedure

The analysis of DPH in blood is performed by
a modification of the methods described by Chang
and Glazko (4) and Evenson et al. (5). To 1 ml of
serum, 0.1 ml of MPPH stock standard is added. The
solution is then made basic with 0.3 ml 0.1 M phosphate buffer, pH 12. The solution is then extracted
with 2 two ml aliquots of diethyl ether. The combined
etheral extracts are washed with 3 ml of 0.1 M phosphate buffer pH 12 and discarded. The basic aqueous
fraction is combined with the phosphate wash, acidified with 1 ml of 10% hydrochloric acid, and extracted with two 5 ml aliquots of chloroform. The
residue, after evaporation of the organic phase, is
dissolved in 15 ul trimethylanilium hydroxide and
then diluted with 50 ul methanol. Samples are then
injected directly into the gas chromatographs (2 ul
samples into the Shimadzu GC and 1 ul samples into
the Finnigan GC-MS).

Both gas chromatographs were equipped with
glass columns (Shimadzu: 6ft x 3mm ID, Finnigan:
5ft x 2mm ID) packed with 3% OV-17 on 80/100 mesh
Chromosorb W HP. During analysis, the column temperature of the Shimadzu GC was maintained at 260°C
and the injection port and detector were maintained
at 310°C. The column and injection port temperatures in the Finnigan GC-MS were 255°C and 300°C
respectively. The flow-rate of the nitrogen carrier
gas in the Shimadzu GC was 30 ml/minute. In the
Finnigan GC-MS operating in the electron impact (EI)
mode, the flow-rate of the helium carrier gas was
8 ml/minute. In the chemical ionization (CI) mode,
the flow-rate of the methane carrier gas was 20 ml/
minute; the source pressure was 1000 microns. The
retention times of the 1,3-dimethyl derivatives of
DPH and MPPH obtained using the Shimadzu GC were 6.1
and 7.8 minutes; 6.0 and 7.7 minutes respectively
using the Finnigan GC-MS in the EI mode. In the CI

mode, the retention times were 3.7 minutes for the
DPH derivative and 4.7 minutes for the MPPH derivative reflecting the faster gas flow rate.

In the EI mode, single ion mass fragmentography was carried out at m/e 194. The mass spectrometer sensitivity was at 10^{-6} amps/volt. The total
ion current was integrated from 35–335 amu with the
mass spectrometer sensitivity at 10^{-7} amps/volt and
the integrator sensitivity at 300. The electron
energy was 70 eV and the analyzer pressure was 10^{-7}
torr. In the CI mode, the mass spectrometer sensitivity was at 10^{-7} amps/volt and the total ion current was integrated from 279–297 amu with the integrator relative sensitivity at 300. The electron
energy was 130 eV and the analyzer pressure was
3×10^{-5} torr using methane as the reagent gas.

RESULTS

Reliability of quantification of DPH by comparing the peak area ratios of DPH and MPPH was demonstrated previously (5, 6). Washing the serum-
basic buffer with ether prior to extraction with
chloroform removed a large fraction of the fatty
acid esters present. These lipids gave rise to appreciable quantities of fatty acid methyl esters by
reaction with TMAH which interfered with the assay
of DPH and MPPH.

For assaying DPH and MPPH in pure solutions
(i.e. plus the contaminants from TMAH), the hydrogen
flame ionization detector (HFID) was set at 3.2 x
10^{-10} amp full scale; the noise level was 1.6 x
10^{-12} amp. 0.1 ug DPH produced a peak 6.1 x 10^{-11}
amp high, with a width at half height of 15 seconds.
The quantity necessary to produce a peak with a 3:1
signal-to-noise ratio was 0.01 ug.

For the total ion current monitoring (EI mode, 35-335 amu) the mass spectrometer was set at 10^{-7} amp/volt; the noise level was 5.5×10^{-10} amp (relative integration sensitivity was set at 300). 0.1 ug DPH produced a peak 8.5×10^{-9} amp high, and a width at half height of 24 seconds. The quantity necessary to produce a peak with a 3:1 signal-to-noise ratio was 0.008 ug.

For the total ion current monitoring (CI mode, 279-297 amu) the mass spectrometer was set at 10^{-7} amp/volt; the noise level was 8.0×10^{-10} amp. 0.1 ug of DPH produced a peak 1.0×10^{-8} amp high, and a half width at half height of 18 seconds. The quantity necessary to produce a peak with a 3:1 signal-to-noise ratio was 0.01 ug. By focusing the mass spectrometer on the quasimolecular ion peak at m/e 281 sensitivity was increased 20 fold; the quantity necessary to produce a peak with a 3:1 signal-to-noise ratio was 0.0005 ug.

With EI mass fragmentography single ion monitoring at m/e 194 the mass spectrometer was set at 10^{-7} amp/volt; the noise level was 5×10^{-10} amp. 0.1 ug of DPH produced a peak 9×10^{-8} amp high, and a width at half height of 24 seconds. The quantity necessary to produce a peak with a 3:1 signal-to-noise ratio was 0.0002 ug.

With clinical samples, the practical lower limit of sensitivity of the HFID was .05 ug (equivalent to 3 ug/ml DPH blood levels). Under these conditions, the hydrogen flame tends to be somewhat noisy and interfering peaks limit sensitivity. Peaks of interest are often masked (or only partially resolved) by biological components present in the sample (Figure 1). In contrast, with GC-MS single ion monitoring at m/e 194, the practical limit of sensitivity using clinical samples remained comparable to that of pure solutions. Quantification down to .0002 ug was readily achieved.

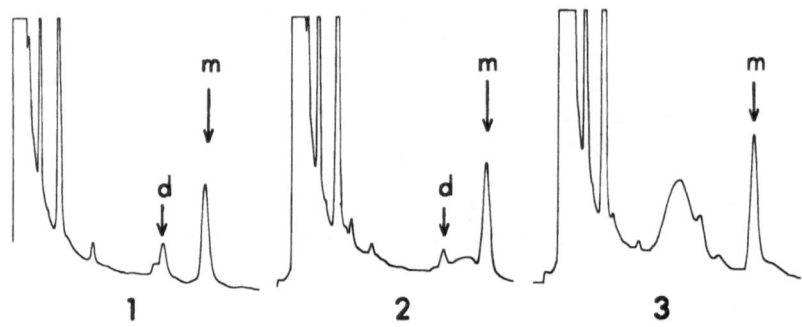

Figure 1. A

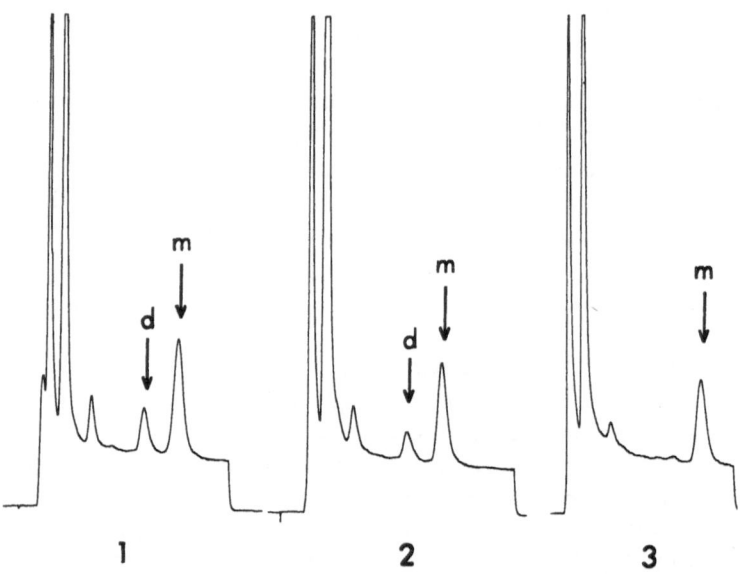

Figure 1. B

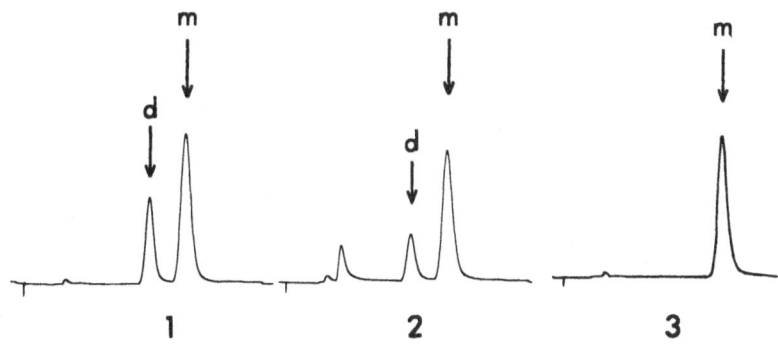

Figure 1. C

Figure 1.:   GLC analyses of 3 clinical samples from
patients receiving diphenylhydantoin therapy.   The
retention times of diphenylhydantoin (d) and the internal standard (m) were 6.1 and 7.8 minutes respectively with HFID; 6.0 and 7.7 minutes with electron
impact GC-MS.   (A) HFID analyses.   (B) GC-MS "total
ion current" analyses of the same specimen.   (C) Mass
fragmentograms of the same specimen with the instrument focused on fragment ion m/e 194.

        A standard curve for the single ion monitoring
of DPH and MPPH at m/e 194 was made by plotting the
ratios of the peak areas, obtained by multiplying
peak height by peak width at half height.   The peak
area ratio of the anticonvulsant to the internal
standard plotted against the actual concentration
gave a straight line passing through the origin
(Figure 2).   The standard curve reflects the therapeutic range of interest in blood level determinations of dilantin (3-30 ug/ml plasma).   However,
proportionality of response to concentration was
maintained for concentrations both above and below
this range.

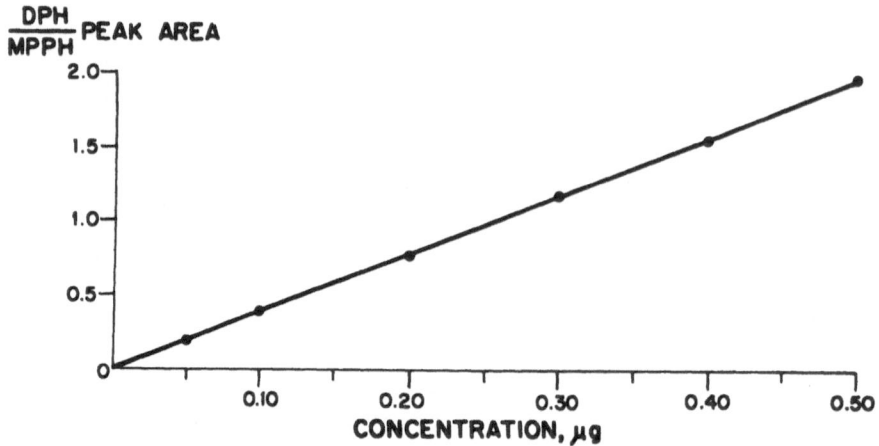

Figure 2.: Standard curve for the quantitative
determination of diphenylhydantoin via single ion
monitoring at m/e 194.

In 20 serum samples there were interferences
in the HFID and the total ion current records that
ranged between the equivalent of 3 and 8 ug of DPH
and MPPH per ml plasma as estimated from the size
of the peaks with retention times close to those of
the two compounds of interest (Figure 1). When
these samples were examined via single ion mass
fragmentography at m/e 194, all interfering components were eliminated and both dilantin and the internal
ternal standard were completely resolved (Figure 1).

DISCUSSION

In recent years, many applications of mass
fragmentography (MF) via multiple ion detection have
been reported, chiefly in the application of qualitative and quantitative analysis of psychoactive
agents, neurotransmitters, and their metabolites and
precursors (1, 2). Rawlins and co-workers (7) reported a specific and sensitive method for deter-

mination of carbamazepine in plasma using 10,11-
dihydrocarbamazepine as an internal standard. The
molecular ions (m/e 236 and 238 respectively) were
monitored and a quantification limit of .05 ug/ml
(using 0.5 ml plasma) was achieved. Strong and
Atkinson Jr. (8) applied similar methods for the
measurements of Lidocaine and its desethylated metabolite in biological samples. Multiple ion monitoring has also been applied for the quantification
of diphenylhydantoin using deuterium-labeled DPH as
the internal standard (9). The 3-monomethylated
DPH derivatives, obtained by treatment with diazo-
methane under mild conditions, were monitored at the
molecular ions (m/e 266 and 276); quantification
down to 0.01 ug/ml (in 0.1 ml plasma) was reported.
However, no supporting data was included to substan-
tiate the claim that only the monomethylated derivatives were obtained during methylation with diazo-
methane. Furthermore, no data was presented that
DPH reacted quantitatively.

A number of studies of MF via single ion monitoring have also been reported. Frigerio and coworkers (10) reported detection of the hallucinogen
STP (2,5-dimethyoxy-4-methylamphetamine) down to
levels of 0.1 ng. In an analogous manner, trace
amounts of pesticides Aldrin (11) and DDT (12) were
detected down to 0.05 and 0.01 ng respectively.
Draffan and co-workers (13) used internal standards
to monitor amylobarbitone and its principal metabolite, 3'-hydroxyamylobarbitone (as their 1,3-dimethyl
derivatives) at m/e 169 and reported a 6:1 signal-to-
noise ratio when injecting 0.1 ng and 0.5 ng respectively.

An internal standard, 5-(p-methylphenyl)-5-
phenylhydantoin was used in the work described here,
to compensate for variability of decomposition,
yield of extraction, and injection volume. Because
of its chemical similarity MPPH is recovered from

plasma and forms a methyl derivative in a comparably
quantitative fashion (4, 14). Both compounds were
chromatographed as their 1,3-dimethyl derivatives
following on-column methylation with trimethylanilium hydroxide (15, 16) in which only one chromatographic peak for each compound appeared.

     Detecting both of these compounds by single ion
monitoring MF requires selection of appropriate ion
of high abundance at a reasonably high mass/charge
ratio so that interference is minimized. The most
abundant ion present in the EI spectra of both DPH
and the internal standard is at m/e 118 (Figure 3).
Although monitoring at m/e 118 will therefore result
in the greatest sensitivity, the relatively low mass

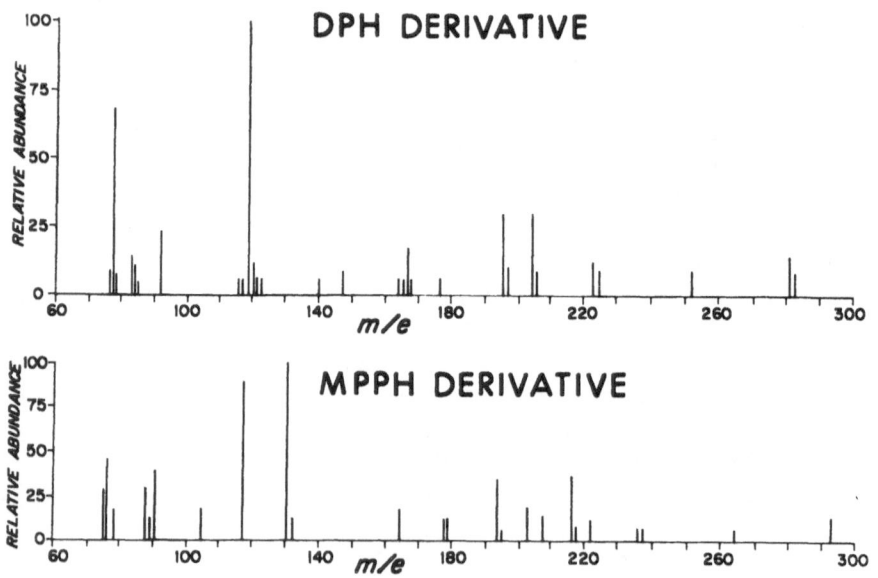

Figure 3.: Electron impact mass spectra of the
1,3-dimethyl derivatives of 5,5 diphenylhydantoin
and of the internal standard, 5-(p-methylphenyl)-5-
phenylhydantoin.

charge ratio of that fragment increases the possibility of interference from other components. Higher
mass/charge ratio ions common to both compounds are
present at m/e 203 and m/e 194. Monitoring of the
latter fragment was undertaken due to its greater
intensity in the MPPH spectrum (Figure 3).

	m/e 194	m/e 203
me-DPH	- 30% of base peak	30% of base peak
me-MPPH	- 35% of base peak	<20% of base peak

There are several options for the use of on-
line mass spectrometry for the analysis of the GC
column effluent: (a) continuous sweep and recording
of the complete mass spectrum in a computer memory
(b) sequential photographic recording of the images
displayed on the oscilloscope (c) multiple ion detection involving time sharing the detector among
several relatively intense ion fragments (e.g.
PROMIM^R-, programmable multiple ion monitor, Finnigan
Corp. (3)) (d) single ion monitoring where the
mass spectrometer is focused on a single fragment
ion.

In conventional electron impact GC-MS, when
the spectrum is scanned, an appreciable fraction of
the time of the detector is occupied scanning regions with either no ions or with ions of relatively low abundance. As a result, sensitivity is comparable to that obtained by conventional HFID
(Figure 1). Using a computer to record spectra in
succession has the same limited sensitivity because
of time sharing. The sensitivity obtained by chemical ionization GC-MS is of the same order of magnitude; essentially the same limitations exist.
The CI mass spectra of these compounds is charac-
terized (when methane is the reagent gas) by $(M+29)^+$
and $(M+41)^+$ peaks indicating the attachment of
$C_2H_5^+$ and $C_3H_5^+$ to the parent ion (Figure 4).

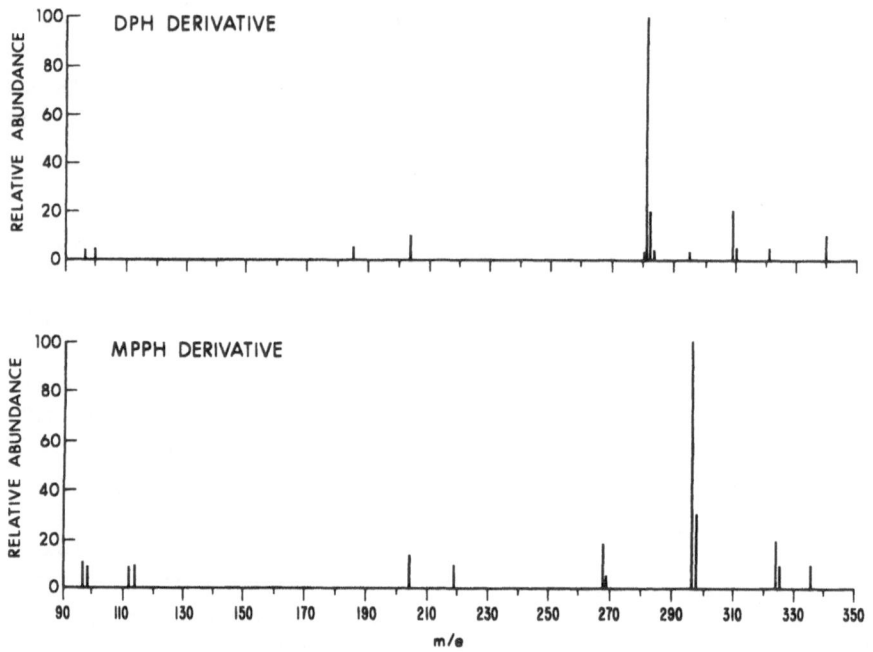

Figure 4.: Chemical ionization mass spectra of the
1,3-dimethyl derivatives of 5,5-diphenylhydantoin
and of the internal standard using methane as the
reagent gas.

     Mass fragmentographic techniques via multiple
ion detection, where the spectrometer is stepped
from one ion to another, improves sensitivity over
conventional GC-MS, although it is somewhat less
specific, because the detector scans only a few relatively intense fragment ions rather than scanning
the entire mass range. Greater sensitivity is avail
able by single ion monitoring because the detector
receives all of the ion fragments of interest generated in the ion source. By focusing on a fixed
mass/charge ratio (instead of scanning) the mass
spectrometer behaves as a highly specific gas chromatographic detector, responding only to compounds
that, on fragmentation, yield ions of the specific

m/e upon which it is focused. No time is spent
scanning between peaks and both chemical noise (e.g.
interfering compounds giving a detector response)
and random statistical noise are reduced relative
to the signal.

Monitoring DPH via single ion mass fragmentography can be performed with high sensitivity and accurate quantification with a minimal decrease in
specificity. By monitoring the EI fragment at m/e
194, diphenylhydantoin was quantified down to 0.2 ng
(with a 3:1 signal-to-noise ratio). 0.5 ng of the
drug could be detected (with a 3:1 signal-to-noise
ratio) by focusing on the quasimolecular ion at m/e
281 in the chemical ionization mode. However, it
is desirable that CI mass fragmentography be carried
out on two fragment ions so that both DPH and the
internal standard can be monitored. The estimated
sensitivity in the CI mode using time sharing between the two quasimolecular ions (assuming half the
ion collecting efficiency of single ion monitoring)
is 1.0 ng (with a 3:1 signal-to-noise ratio). Thus,
single ion monitoring MF in the EI mode offers a 5
fold increase in sensitivity. The advantages of CI
time sharing MF (over EI single ion monitoring MF)
are observed in low resolution systems involving direct injection or very rapid GC elution where simplicity of the CI spectra may allow detection of the
compound of interest in complicated matrices.

Greater total sensitivity might be available
if the total ions formed were continuously recorded-
as for example, if the ion current deflected to the
rods of the quadrupole were recorded.

SUMMARY

A method is described for analysis of diphenylhydantoin in clinical serum samples employing:

1. addition of a dilantin analogue as internal
standard 2. extraction by a procedure that selectively removes some of the interfering lipids 3.
analysis by GC-MS mass fragmentography in which an
ion is monitored that is common to both diphenylhydantoin and its analogue. The sensitivity is such
that .0002 ug produces a peak at 3 times the noise
level. The spectrum is free from interferences at
sensitivity levels at which standard HFID records
interfering peaks equivalent to 3-8 ug/ml and total
ion monitoring records interfering peaks equivalent
to 3-8 ug/ml of diphenylhydantoin. Using the common
ion approach, EI gives greater sensitivity (at 3x
noise signal) by a factor of 5 versus estimated CI
sensitivity by time sharing. The use of CI time
sharing MF may be advantageous in low resolution
systems.

REFERENCES

1. A.E. Gordon and A. Frigerio, J. Chromatog., 73,
    401 (1972).
2. D.J. Jenden and A.K. Cho, Ann. Rev. Pharmacol.,
    13, 371, (1973).
3. J.M. Strong and A.J. Atkinson Jr., Finnigan
    Spectra, 4, #3, Dec. 1974.
4. T. Chang and A.J. Glazko, J. Lab. Clin. Med.,
    75, 145 (1970).
5. M.A. Evenson, P. Jones, and B. Darcey, Clin.
    Chem., 16, 107 (1970).
6. E.M. Baylis, D.E. Fry, V. Marks, Clinica Chimica
    Acta, 30, 93 (1970).
7. L. Palmer, L. Bertilsson, P. Collstei, and M.
    Rawlins, Clin. Pharm. Thera., 14, 827 (1973).
8. J. M. Strong and A.J. Atkinson, Jr., Anal. Chem.
    44, 2287 (1972).
9. R. Rane, M. Garle, O. Borga, and F. Sjoqvist,
    Clin. Pharm. Thera., 15, 39 (1974).

10. A. Frigerio, R. Fanelli, and B. Danieli,
     Chem. & Ind., 19, 769 (1972).
11. E.J. Bonelli, J.B. Knight, and M.S. Story,
     Finnigan Application Tips, #24 (1971).
12. Application Note #4, Varian Mat, (GmbH) Sept.
     1971.
13. G.H. Draffan, R.A. Clare, and F.M. Williams,
     J. Chromatog., 75, 45 (1973).
14. J. H. Goudie and D. Burnett, Clinica Chimica
     Acta, 43, 423 (1973).
15. E. Brockmann-Hannsen and T. Oki, J. Pharm. Sci.,
     58, 370 (1969).
16. H.J. Kupeferberg, Clinica Chimica Acta, 29, 283,
     (1970).

ACKNOWLEDGEMENTS

This work was supported by N.I.H. Grant GM 19478.

NALYTICAL APPLICATIONS OF A NEW SPECTROSCOPIC TOOL:

OHERENT ANTI-STOKES RAMAN SPECTROSCOPY (CARS)

A. B. Harvey, J. R. McDonald, W. M. Tolles†

Physical Chemistry Branch (Code 6110)

Naval Research Laboratory
Washington, D. C.   20375

Conventional, spontaneous Raman spectroscopy has been
n extremely useful technique for analytical applications
1-5). It is used for both qualitative and quantitative
nalysis in cases which complement infrared methods. For
xample, homonuclear diatomic molecules, H_2, D_2, O_2, N_2,
I_2, etc. cannot be detected by infrared spectroscopy.
oreover, certain "symmetrical" vibrational modes are
ften only weakly allowed in the infrared spectrum, but
uch vibrations are frequently very intense in the Raman
ffect. An example is the vibrational spectrum of Thiokol
ubber (6). The ir spectrum of this material shows the
resence of C-O-C linkage but no indication of C-S or S-S
onds. The Raman spectrum exhibits no suggestion of
ther-like moieties but very strongly displays the intense
-S and S-S vibrations. Hence, both spectra are necessary
o identify the material. Raman spectroscopy has certain
dvantages over infrared spectroscopy, especially with
espect to sample handling. With Raman spectroscopy, one
an obtain results from samples in sealed tubes, whereas
areful sample preparation is often required for infrared
pectroscopy. Raman spectra of aqueous solutions are much
asier to obtain since Raman scattering in water is rela-
ively weak, whereas infrared absorption in water is
early ubiquitous. Thus, Raman spectroscopy is a potent
ool for the bio-related sciences. Raman spectroscopy is

Dept. of Chemistry and Physics
Naval Postgraduate School
Monterey, California 93940

also becoming a very useful technique as a diagnostic tool for combustion and aerodynamic applications (7-12).

Despite these advantages, Raman spectroscopy suffers from some serious drawbacks. First, the Raman effect is a very weak phenomenon, typically one photon scattered for every $10^8 - 10^{10}$ exciting photons. Second, collection efficiencies of the scattered photons are also small, since Raman scattering takes place in all directions (for unpolarized light). Third, laser-induced fluorescence is often more intense than Raman scattering and is therefore a very serious interference.

Some of the drawbacks of ordinary Raman spectroscopy can be overcome by a new Raman technique called Coherent Anti-Stokes Raman Spectroscopy or CARS. The effect was first observed by Terhune (13) over a decade ago as part of stimulated Raman emission, and its progress as an analytical tool was retarded, mainly because of inadequate tuning devices. Nonetheless, several research efforts utilizing the phenomenon for measuring resonances (14-18), nonlinear coefficients and properties (19-27), kinetics (28, 29), etc. have appeared in the literature over the past few years and only recently has the technique been exploited as a general spectroscopic or analytical method (30-43). This new method is actually one of several other quite different nonlinear Raman effects which have appeared recently. All of these methods, including improvements in normal Raman, have come about by the advent of the laser. Table I lists these techniques. Because these phenomena are often confused or not clearly understood, we shall try to briefly describe each and, by way of comparison, relate their utility and promise for analytical applications.

Table I

    Conventional Spontaneous Raman Spectroscopy
    Resonance Raman Effect
    Inverse Raman Effect
    Hyper-Raman Spectroscopy
    Raman-Induced Kerr Effect
    Coherent Anti-Stokes Raman Spectroscopy (CARS)

CONVENTIONAL SPONTANEOUS RAMAN
AND STIMULATED RAMAN EFFECTS

These two phenomena will be discussed collectively since normal Raman spectroscopy is a special case of the general theory on stimulated Raman effect, a good development of which may be found in a discussion by Lallemand (44). Normal Raman is a spontaneous, inelastic scattering phenomenon which for Stokes emission results in a red shifted photon and a molecule which has been promoted to an excited state. For anti-Stokes emission the scattering is the result of an interaction with molecular species which are already excited, thus giving rise to a de-excited molecule and a blue shifted photon. A trace of the frequency shifts is a record of the (vibrational) Raman spectrum of the material. For very intense laser sources, the spontaneously generated Stokes photons begin to build up gain in the direction of laser propagation through the medium. The intensity of this Raman emission has exponential growth with laser intensity. Hence, it is clear why stimulated Raman was not observed before intense lasers were discovered.

Conversion efficiencies of laser power into stimulated Raman emission can be very impressive, as much as 20%. Since one obtains such intense Raman signals, one might ask why stimulated Raman effect would not replace normal Raman spectroscopy which is many orders of magnitude less intense. The reasons are quite simple. Stimulated Raman spectroscopy is not nearly as general. Not every species nor every molecular mode may be made to go stimulated. Usually, only the most intense, narrow lines (the totally symmetric vibrations) are observed. This is because the exponential growth in the few intense modes will simply convert more laser power into those modes going stimulated rather than bringing about stimulated emission in other modes. Moreover, losses in transmission, e. g., the simultaneous generation of anti-Stokes emission, cause the exponential dependence to take on a threshold character. Hence, stimulated Raman does not appear at all until the number density and laser power are large enough and even then, if one is near threshold, large fluctuations in stimulated emission occur with slight variations in laser amplitude. Thus, only a few gases have been made to produce stimulated Raman emission, and the gases must be at

many atmospheres pressure. It is therefore clear that
this method is not useful for general analytical applications.

RESONANCE RAMAN EFFECT

As mentioned earlier, normal Raman effect is a very
weak phenomenon which unfortunately affects its usefulness
as a sensitive analytical tool. Therefore, any means to
increase the intensity of normal Raman scattering is
sought. One method for specific substances is Resonance
Raman Effect. For materials which have strong (electronic)
absorption in the region of the (laser) excitation, Raman
scattering of certain vibrational modes may increase mani-
fold as the excitation frequency approaches that of the
electronic resonance. Enhancements of up to 10^6 can be
achieved. However, the technique is useful only for substances that absorb in the region of excitation. In this
manner, radical ions (45), ions in aqueous solution (46),
gases (47), etc. have been observed to exhibit strong
resonance effects. In addition to increasing the sensitivity of normal Raman spectroscopy, the Resonance Raman
Effect can provide useful structural and spectroscopic
information.

INVERSE RAMAN EFFECT

Inverse Raman Effect was discovered by Jones and
Stoicheff in 1964 (48), and the method has not been well
explored. Basically, the method involves the use of two
lasers - one fixed and monochromatic, while the other may
be broad in frequency or narrow and tunable. Both beams
are propagated through the medium under investigation, and
absorption of the short wavelength laser at frequencies
which correspond to the anti-Stokes resonances is monitored. Some research in liquids (49-54) and very few
studies in gases (55) have been performed. However, one
needs to use very high-power lasers in order to achieve
appreciable absorption. In fact, usually laser powers
just short of effecting stimulated emission are used.
The method appears to be useful for obtaining spectra of
condensed media where absorptions can be measured easily
in the presence of single shot or averaged pulsed signals.
For low concentrations or for low number densities, as in

gases, the absorptions become quite small, and it becomes
very difficult to observe small depletions in (dye) laser
power. Because one is measuring absorption in the laser
of shorter wavelength, inverse Raman, like CARS, is not
usually hindered by fluorescence in the sample.

HYPER-RAMAN SPECTROSCOPY

Hyper-Raman Spectroscopy is another kind of nonlinear
effect, also called inelastic second harmonic scattering,
and occurs as Stokes shifted photons with respect to the
second harmonic of the exciting laser frequency (56). The
efficiency of this process is very low, even by comparison
with normal Raman spectroscopy. Gases, for instance, must
be examined at very high pressures (several tens of atmospheres) even with megawatts of laser power. At these high
pressures and laser powers, only a few photons are emitted
per laser pulse. Unlike stimulated Raman, CARS, and inverse Raman, hyper-Raman scattering takes place over
all solid angles, much like normal Raman spectroscopy.
Although the method appears to have little use in most
analytical applications, it can be valuable in structural
determinations and in vibrational analysis since selection rules are quite different from normal Raman spectroscopy.

RAMAN-INDUCED KERR EFFECT

Raman-Induced Kerr Effect is a newly discovered technique (57) and, as such, its potential for analytical
application can only be estimated. Like CARS, the method
is a nonlinear phenomenon which utilizes two lasers; one
is narrow band and fixed, while the other may be broad
band and fixed or narrow band and tunable. If the second
laser is broad band, then photons created by the process
must be separated by a monochromator. Both beams must be
highly polarized, focused and overlapped in the medium.
The broad band or the tunable source is called the probe
beam, while the other laser is called the pump beam. The
probe beam intensity is first blocked by an analyzer
(polarizer oriented perpendicular to the direction of
polarization of the probe beam) after the sample cell in
the absence of the pump beam. When the pump beam is

admitted, transmission of the probe beam through the analyzer at both Stokes and anti-Stokes frequencies corresponding to the difference in frequency between the two
laser beams is brought about by the Raman-Induced Kerr
Effect. If the pump beam is circularly polarized, background transmission, caused by the usual (nonresonant)
Kerr effect, is minimized. Like CARS, only a few kilo-
watts of laser power in each beam are necessary to observe
the effect in liquids and solids and much higher power
will be necessary to observe gases. In principle, the
method appears to have an advantage over CARS in that nonresonant background interference can be eliminated. In
practice this depends on how efficiently the lasers can
be polarized and the rejection ratio of the polarizers.
Its ability to provide trace analysis is highly dependent
on these factors. Also, it is quite possible that turbu-
lence, hot spots and similar effects in discharges,
plasmas, sparks, etc. may cause anisotropies which more
severly affect this method than normal Raman or CARS. At
this early stage, it is clear that the method is potentially useful for many applications and warrants careful
examination in the future.

COHERENT ANTI-STOKES RAMAN SPECTROSCOPY

We have chosen to discuss this topic last, since it
is the main subject of this presentation and since Coherent Anti-Stokes Raman Spectroscopy probably has more
general utility than all the other methods except normal
Raman and possibly Raman-Induced Kerr Effect. The technique often generates high conversion efficiencies (as
much as 1%) which are many orders of magnitude greater
than normal Raman (31-33, 41-43). CARS has no threshold
and can be a practical tool at almost all levels of laser
power and number densities. Unlike normal Raman, CARS
signals are generated in the form of a coherent laser-
like beam in the anti-Stokes region. Thus, collection
efficiencies and discrimination against interferences such
as fluorescence are extremely high. All transitions
active in normal Raman are CARS active. Except where
background generation interferes, the CARS method can be
used to detect moderate to low concentrations, especially
in gases, but is not yet a good method for trace analysis
because of problems with background signal generation.

However, there appear to be methods whereby this limitation may be reduced or eliminated.

Basically, the experimental arrangement for CARS experiments can be quite simple. Such an arrangement is described in Figure 1. Two lasers must be employed, usually one fixed, the other tunable. The tunable laser is often a narrow band dye laser, although a broad band laser and monochromator may also be used. However, there are many variations, such as two synchronized flash pumped dye lasers, two nitrogen laser pumped dye lasers, etc. Anti-Stokes emission (at $\omega_{as} = 2\omega_\ell - \omega_s$) is generated when two photons of the higher frequency laser are mixed with one photon of the lower frequency laser.

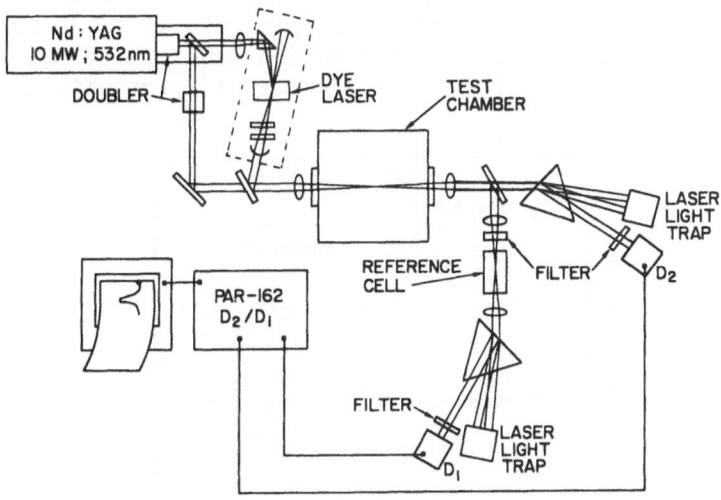

Fig. 1 An apparatus used at NRL for generating CARS spectra of gases. The reference cell is not necessary, but since the CARS signals are strongly dependent on changes in laser amplitude, CARS conversion in a high-pressure nonresonant gas (Ar) is very beneficial in reducing noise or averaging time. The Nd:YAG laser has a repitition rate of 10 pps, a line width of ~ 0.05 cm^{-1} and ~ 20 nsec. The dye laser (~ 0.6 MW) has a line width of ~ 0.3 cm^{-1} using a single grating and an Inchworm (T.M.) driver. The insertion of an etalon can reduce the dye laser width to that comparable to the Nd:YAG laser.

A CARS spectrum of D_2 gas taken with two Chromatix - CMX-4 flash pumped dye lasers is illustrated in Figure 2.

The CARS emission takes place because the material under investigation has a polarization which is nonlinear with respect to field strength. The polarization can therefore be expressed as a series expansion in simplistic form as follows:

$$P = \chi^{(1)} E + \chi^{(2)} E^2 + \chi^{(3)} E^3 \dots ,$$

where P is the polarization, E is the electric field strength, and χ^i is the ith coefficient (or susceptibility). The first term in the equation above is the linear response to the medium, whereas the subsequent terms are responsible for the higher order effects. For example, the first nonlinear coefficient, $\chi^{(2)}$, is responsible for second harmonic generation and for the hyper-Raman effect. The third-order term, $\chi^{(3)}$, results in

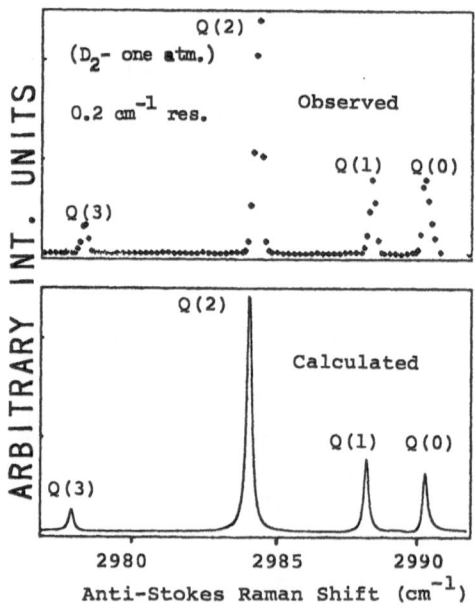

Fig. 2 Observed and calculated spectra of D_2 by CARS utilizing two Chromatix CMX-4 tunable dye lasers (1-3 kw, 0.1 cm^{-1} and 1 μsec width in each beam).

third harmonic generation and other effects such as CARS.
It should be emphasized that this relation is simplified.
For example, there are usually integers which appear as
multiplicative factors before each of the nonlinear coefficients which are tensors. In the CARS process this
factor is three, which some authors include in deriva-
tions, while others omit it (19). Throughout our discussion we shall separate out this factor of three; hence,
our susceptibility will be one-third that of some authors
(39, 41-43). It is quite clear that if the field strength
is small, these higher-order phenomena are not likely to
be very efficient, since they will not appreciably contribute to the polarization. However, for high values of
E, as in the case of high-power lasers, the third-order
contribution might be quite high, even though $\chi^{(3)}$ itself
is small. Thus, CARS is most efficient for high-power
lasers.

We shall not attempt to develop the basic theory for
CARS but simply state the results and refer the reader to
a series of references (14, 19, 58). The efficiency of a
CARS signal is defined and determined according to the
following equation:

$$\epsilon = \frac{P(\omega_{as})}{P(\omega_s)} = \left(\frac{12\pi^2\omega_{as}}{c^2 n_{as}}\right)^2 |\chi^{(3)}|^2 I_l^2 L^2 \tag{1}$$

where ϵ is the efficiency of the CARS signal production,
$P(\omega_{as})$ is the power at the angular anti-Stokes frequency,
$P(\omega_s)$ is the power at the Stokes angular frequency, n_{as} is
the index of refraction at ω_{as}, $\chi^{(3)}$ is the third-order
molecular susceptibility, I_l is the intensity at ω_l and L
is the path length through which the CARS signal is generated. Equation (1) is valid only for unfocused plane
waves. For focused beams, where one assumes the focal
region to be a cylindrical volume of length equal to the
confocal parameter and radius equal to that of the beam
waist radius of a guassian beam, then it can be shown (43)
that the efficiency becomes:

$$\epsilon = \left(\frac{48\,\pi^3}{\lambda_a \lambda_{as} n_{as} c}\right)^2 \left|\chi^{(3)}\right|^2 P_i^2 \qquad (2)$$

where λ is wavelength. This model for focused beams is not quite correct since it assumes plane waves and a truncated mixing volume. A more precise approach to focusing has been carried out elsewhere (40, 59, 60). Nonetheless, equation (2) is useful for most purposes. The third-order, nonlinear coefficient or susceptibility is calculated to be:

$$\chi^{(3)} \approx \frac{2Nc^4}{3\hbar\omega_s^4}\left(\frac{d\sigma}{d\Omega}\right)\left[\frac{1}{2\Delta\omega - i\Gamma}\right] \qquad (3)$$

where N is the number density, $(d\sigma/d\Omega)$ is the normal Raman differential cross section, $\Delta\omega = (\omega_\ell - \omega_s) - \omega_V$, ω_V is the resonance frequency, and Γ is the normal Raman line width (FWHM). Thus, $\chi^{(3)}$ is a complex number, composed of a real part, χ', and an imaginary part, χ''. It is the square of the absolute value (or modulus) of $\chi^{(3)}$ which is related to the CARS emission intensity or $|\chi^{(3)}|^2 = \chi \cdot \chi* = (\chi')^2 + (\chi'')^2$, where $\chi*$ is the complex conjugate of χ. In Figure 3 we plot the real and imaginary parts of the susceptibility, and in Figure 4 we plot their squares. Note that when $\Delta\omega = 0$ of equation (3), i.e., directly on resonance, $\chi' = 0$ and χ'' becomes:

$$\chi^{(3)} = \chi'' = \frac{2Nc^4}{3\hbar\omega_s^4\Gamma}\left(\frac{d\sigma}{d\Omega}\right) \qquad (4)$$

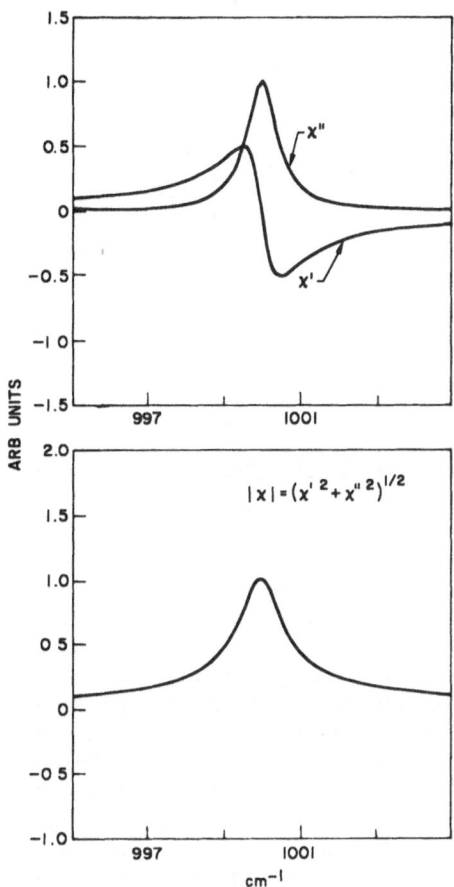

Fig. 3 Calculated shapes of the third-order nonlinear
susceptibility and its real and imaginary components. A
normal Raman line width (FWHM) of 1 cm^{-1} was assumed. The
resonance is assumed to be centered at 1000 cm^{-1} shift.

At the peak of the resonance then, the CARS signal is
proportional to the square of the normal Raman cross section and varies inversely as the square of the normal
Raman line width. Note also that according to equations
(1-4), the power in the CARS beam is dependent on the cube
of the overall laser power and the square of the number
density in contrast with normal Raman which varies linearly
with both.

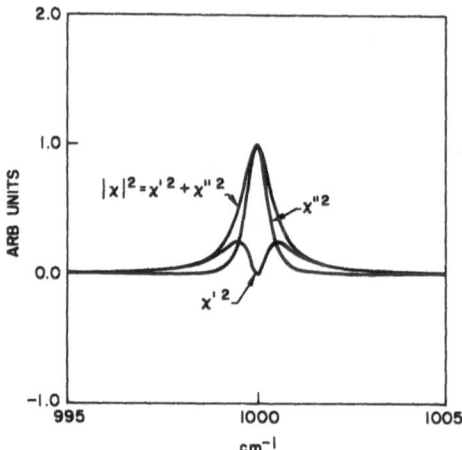

Fig. 4 The squares of the functions plotted in Fig. 3 are displayed here.

        It should be mentioned that the efficiencies calcu-
lated in equations (1 and 2) are valid only for the phase-
matched condition, that is, the three waves ω_{as}, ω_{ℓ}, and
ω_s gradually slip out of phase because they move at different velocities through the medium (dispersion in refractive index). Figure 5 shows how the wave vectors associated with each wave can be crossed to achieve exact phase
matching. For most cases, this crossing angle is usually
small ($\sim 1°$ for benzene at $\Delta\nu = 992$ cm^{-1}). For the
collinear case equation (1) must be replaced by the
following relation which shows a periodic variation in
efficiency with path length (see Figure 6 and reference
(16)).

$$\epsilon = \left(\frac{12\pi^2\omega_{as}}{c^2 n_{as}}\right)^2 |\chi^{(3)}|^2 I_\ell^2 \left(\frac{2L_c}{\pi}\right)^2 \sin^2\left(\frac{L}{L_c}\cdot\frac{\pi}{2}\right)$$

At the first maximum the path length we define as the
coherence length, L_c, the path length beyond which the
process begins to decrease in conversion efficiency by
destructive interference as the three waves (ω_S, ω_ℓ, and
ω_{as}) gradually slip out of phase because of dispersion.

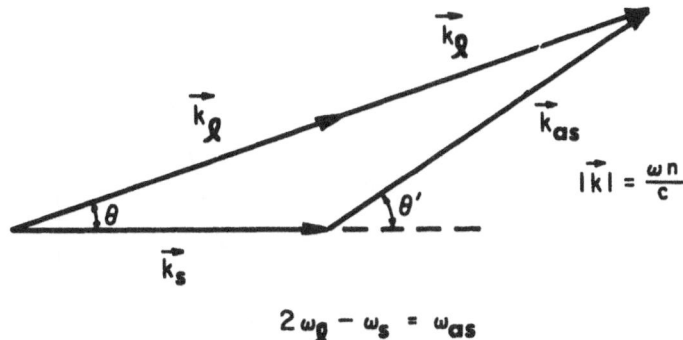

$$2\omega_{\ell} - \omega_s = \omega_{as}$$

Fig. 5 Phase-matching condition for CARS. In condensed media the laser beams must be crossed by angle θ in order to achieve maximum CARS conversion. The anti-Stokes radiation is emitted at different angle θ' which aids in spatial filtering of the CARS beam from the laser beams. Note that for gases, crossing the beams is usually not required since dispersion in gases is very small.

Throughout the previous arguments we have discussed only the resonant contribution to the susceptibility. However, there is also a nonresonant part, χ^{nr}, which is due to the electrons in the medium. We shall also take this quantity to include residual resonant susceptibility from distant resonances. Thus, the total susceptibility may be written as:

$$\chi^{(3)} = (\chi' + \chi^{nr}) + \chi''i, \tag{5}$$

where χ' and χ^{nr} are grouped together because they are both real quantities. As mentioned earlier, the CARS signal is proportional to the square of the absolute value of $\chi^{(3)}$:

$$|\chi^{(3)}|^2 = (\chi' + \chi^{nr})^2 + (\chi'')^2 \quad \text{or}$$
$$= (\chi')^2 + (\chi^{nr})^2 + (\chi'')^2 + 2\,\chi'\cdot\chi^{nr}, \tag{6}$$

For small or nearly zero values of χ^{nr}, we can neglect the $\chi'\cdot\chi^{nr}$ cross term in the equation above, and nothing unusual develops. However, for values of χ^{nr} which are not negligible, then $(\chi')^2$ and $(\chi'')^2$ are small, leaving only the $2\,\chi'\cdot\chi^{nr}$ term (the $(\chi^{nr})^2$ is assumed to be a constant). This can occur for small concentrations of a

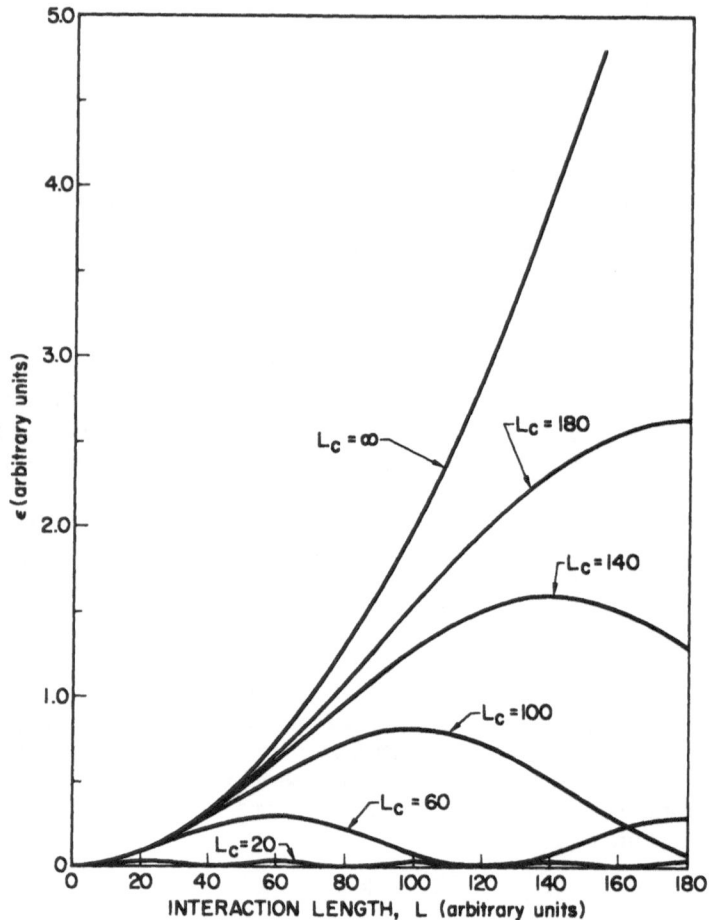

Fig. 6 CARS conversion efficiency as a function of interaction length for collinear beams. Note the maximum conversion efficiency for the nonphase matched cases.

solute with respect to the solvent or for very weak
resonant components. In this case, the resulting line
shape will not be symmetrical but will take on the un-
symmetrical shape of the real part of the susceptibility,
χ'. Moreover, the peak of the resonance will no longer
be centered at the maximum of the line. These effects
are illustrated in Figure 7. The most serious drawback
of these interferences is the resulting limit on trace

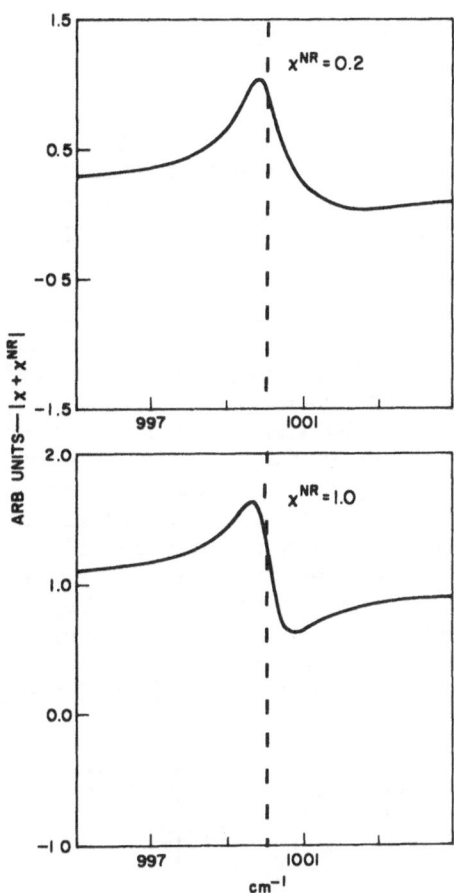

Fig. 7   The affects of mixing 20% (upper trace) and 100% (lower trace) nonresonant susceptibility on the absolute value of the total susceptibility.   Note the shift in peak height and the gradual change in band shape to that of χ'. As before, a line width of 1 cm^{-1} and shift of 1000 cm^{-1} for the calculated curves are assumed.

analysis, since one can observe CARS to concentration levels not much lower than the background generated in the solvent will allow.   Because the strong dependence of line width, gases are much less hindered by background generation.   For example, H_2 was detected in N_2 to 10 ppm (41), whereas sodium benzoate in aqueous solution could not be detected below 0.5 weight % (33).   For undiluted or mildly diluted gases, CARS has tremendous potential for detecting species at low pressures ($<$ 1 millitorr).

As indicated by equation (6), when the χ^{nr} becomes appreciable, then one approaches the linear regime where the signal is linearly dependent on number density and normal Raman cross section. As has been noted previously (32), spectral regions of complex overlapping bands have CARS spectra which very remarkably resemble normal Raman spectra. Moreover, CARS should exhibit somewhat better effective resolution than normal Raman spectroscopy. A comparison of the two techniques can be made because conventional Raman spectroscopy can be related to the imaginary part of the susceptibility, χ'' (44), whereas CARS is brought about by the contribution from both the real and the imaginary parts. In the case of overlapping resonances, the real parts of each resonance combine in such a way as to lower the total susceptibility between the bands, thus giving rise to improved line separation. Synthetically drawn curves in Figure 8 illustrate these effects quite well, and experimentally they may have already been observed (44). Thus, we may further generalize by saying that normal Raman is related to the sum of resonances via χ'', whereas CARS is determined by the square of the sum or:

$$I_{normal\ Raman} \propto \sum_i (\chi_i'')$$

$$I_{CARS} \propto |\sum_i \chi_i^{(3)}|^2, \tag{7}$$

Thus, it is clear that CARS is quite different from normal Raman in that it exhibits strong interaction with neighboring resonances and the background. It is a spectroscopic tool which reflects not only the microscopic properties of the medium (molecular energy levels) but also the bulk properties, i. e., index of refraction. Thus, two neighboring vibrational frequencies interact by means of cross term in equation (7), whether the frequencies are part of one molecule or that of two molecular species in a mixture.

One of the most important advantages of CARS is the coherency and the anti-Stokes nature of its emission which strongly discriminates against the interferences from spontaneous emission, especially laser-induced fluorescence, which so frequently obscures normal Raman spectra (natural products, bio-related materials, and commercial polymers, in particular). A dramatic illustration of this capability has been demonstrated by the addition of a fluorescent dye (Rhodamine 6G) to a sample of benzene (33). Despite the very strong fluorescence generated in the benzene, CARS spectra were very easily observed. Another important example of the potential of this new tool is its ability to probe plasmas, discharges, sparks, etc., where normal Raman would be overwhelmed by spontaneous emission. We have recently recorded the CARS spectrum of vibrationally hot $(v = 1)$ and ground state $(v = 0)$ D_2 resident in the middle of an electrical discharge region (61). At 48 torr the signals from D_2 were strong enough to easily record on color film. From the relative intensities of the Q branches, both rotational and vibrational temperatures can be measured. It should be mentioned that because the two exciting laser beams can be focused to a very small volume $(\sim 0.001 \text{ mm}^3)$, CARS can be used to construct three-dimensional temperature and concentration maps of flames and other combustion systems. This has been accomplished by Taran and coworkers (40-43).

Although CARS efficiencies are greatest for intense, pulsed laser sources, the technique can be carried out using cw lasers. Barrett and Begley (30) have recently detected methane (1 atm) using low-power cw lasers. For low-level detection, however, cw sources are probably not powerful enough to be practical, despite their inherently better stability and beam quality.

As mentioned earlier, one of the most severe limitations associated with CARS as an analytical tool for trace detection is the interference of background generation. There are a few suggestions which may, in certain cases, reduce this limitation - one idea is to create resonance enhancement of the CARS signal over the background by the encroachment of the CARS signal or the laser pump frequency upon an electronic resonance, similar to resonance Raman effect. Such enhancements in CARS have already been observed (62, 63). Another method of increasing CARS signals above the background for gases may be the use of rotational transitions which are both more intense and narrower in line width, both factors which favor CARS efficiencies. However, this technique has several drawbacks which include the difficulty in generating and

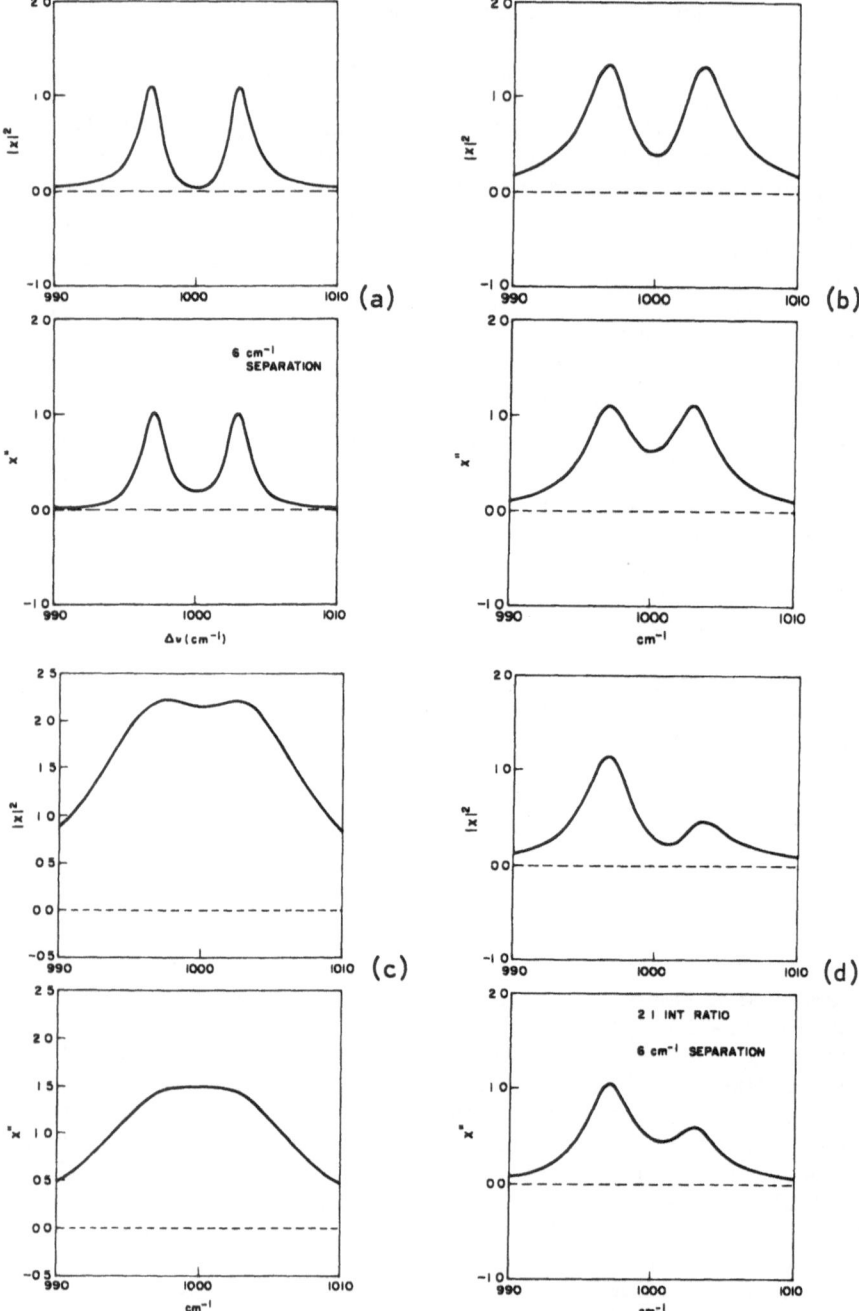

separating lasers similar in frequency, utility restricted to only simple molecular species, and a decrease in CARS efficiency at high temperatures since CARS signals are dependent on the difference in number density between levels involved in a particular transition (39, 58). There are other ideas to reduce or eliminate the background, but it is too early to discuss them. However, there is hope that these new ideas may significantly reduce or even eliminate background interferences (see, for example, reference (64)).

For completion we should mention that coherent emission is generated at the corresponding Stokes frequency as well as the anti-Stokes region, i. e., $\omega_{as} = 2\omega_{\ell} - \omega_s$, $\omega_{Stokes} = 2\omega_s - \omega_{\ell}$. If the intensity at ω_s is much lower than ω_{ℓ}, then the Stokes emission will be much lower than the anti-Stokes output (CARS). Even though the Stokes generation will have fluorescence interferences, there may be times when it is advantageous to search for Stokes emission (e. g., when the CARS signal is absorbed).

Before we end this discussion, we should mention costs. Of all these Raman methods, only normal Raman spectroscopy has developed into the commercial market. The other methods are either too costly, relatively new

---

Fig. 8 Synthetically drawn traces of two resonances separated by 6.0 cm^{-1} with varying bandwidths (a) 2.0 cm^{-1}, (b) 4.0 cm^{-1}, (c) 10.0 cm^{-1} and (d) 4.0 cm^{-1} 2:1 peak height. Upper curves of each are plots of $|\chi|^2$ or CARS signal. Lower curves show respective plots of χ'' or normal Raman signal. Clearly, CARS yields slightly better band separation because of the negative contribution of the real component of susceptibility.

and yet developing, not very general in their utility or
some combination of the above, so that commercial development of the techniques will await some time in the future.
However, it is curious that all of these methods utilize
much the same equipment, i. e., a high-power, pulsed laser
(e.g., Nd:YAG at 532 nm), possibly a second tunable or
broad band (dye) laser, possibly a monochromator, detectors and miscellaneous optics (polarizers, reflectors,
beam splitters, lenses, filters, etc.), the major invest-
ment being the high-power laser sources which, in
themselves, may cost more than a routine, commercial Raman
apparatus complete with cw laser. Of course, these costs
of material and capital equipment do not include maintenance or salaries of specialists to construct and operate
the apparatus.

SUMMARY

        In the discussion above, we have attempted to briefly
describe a number of different Raman methods which may be
easily confused and misunderstood, and we have attempted
to assess the ability or promise of each to provide analytical information. However, it has been the main purpose of this paper to describe a relatively new Raman
method (CARS) in more detail. Basically, because of the
high-conversion efficiencies, the coherency of the CARS
emission and the discrimination against fluorescence,
CARS appear to be a superior tool for combustion, plasma,
and laser diagnostics. It will also be a developing
spectroscopic method for the bio-related sciences and
polymer industry. It will, undoubtedly, be a technique
for studying picosecond processes, fluctuation phenomena,
and photolysis. For conducting high-resolution Raman
spectroscopy, CARS is clearly unsurpassed. However,
because of the costs involved and the problems with
tunable sources which are still developing, it will be
some time before the method will be considered routine.
In Table II, we attempt to summarize these conclusions.

        As a final note, we should like to summarize the
advantages and disadvantages of CARS over normal Raman
spectroscopy for specific analytical applications. This
summary appears in Table III.

TABLE II

	Raman Efficiency	Discrimination Against Fluorescence	Trace Analysis Capability	Det. of Low No. Densities	Comments
Normal Raman*	low	low	moderate-low	mod-low (100 mtorr)	General application.
Stimulated Raman	high	high	very low	very low (20 atm)	Limited usefulness.
Inverse Raman	high	high	low	low (? 10 torr)	Good for condensed phase. Not sens. for gas anal. at low pressure.
Hyper-Raman	very low	high	very low	very low ($\sim$1 atm)	Sens. very low, good for some structural anal.
Raman-Induced Kerr Effect	high	high	moderate (?)	moderate (?)	Too early for good assessment but has definite potential.
CARS*	high	very high	low (at present**)	high (< 1 mtorr)	Excel. for hi resolution and anal. of gases at low pressure. Problems with trace anal. (low ppm)

* These assessments do not include improvements effected by electronic resonance enhancements.

** The limit to CARS for trace analysis is the interference from background generation of solvent or diluent gas. At this writing, there are a few ideas not yet fully tested which may markedly reduce or eliminate background generation.

Table III

ADVANTAGES OF CARS

1. High Raman conversion efficiencies (as much as 1% or
    $>10^5$ greater than normal Raman conversion).

2. Higher collection efficiencies because CARS is gener-
    ated in a laser-like beam.

3. Monochromator not necessary, spectral width determined
    by the laser line width.

4. Moderately high resolution (~ 0.03 cm^{-1}) is routine by
    CARS because lasers of such line widths are readily
    attainable (commercially). Very high resolution
    ($<10^{-3}$ cm^{-1}) at modestly high powers (50 kw) is feas-
    ible with lasers already devised (65).

5. Laser-induced fluorescence and spontaneous emission
    from flames, plasmas, discharges, etc. are usually
    not interferences by CARS because of the coherency
    and spectral properties of the phenomenon.

6. High sensitivity for gases at low pressures (calcula-
    ted to be $\sim 10^{-10}$ atm for very high-powered lasers
    (43)).

7. Because of the higher conversion efficiencies and the
    need to use pulsed sources, new and interesting ex-
    periments in photochemistry, kinetics and molecular
    relaxation and fluid dynamics seem possible.

8. Since CARS is a process involving four waves, there
    appears to be more information potentially available
    in measuring polarization ratios. Hence, it may be
    possible to make better assignments of normal modes
    to their correct irreducible representations.

DISADVANTAGES OF CARS

1. One of the most serious problems with CARS as an
    analytical tool appears to be the generation of back-
    ground radiation due to the nonresonant part of the

susceptibility which limits, at present, detection
to ~ 1% for aqueous solution and about 10 ppm in
gases. As mentioned in Table II, this problem may be
solved in the near future.

2. The method is probably not useful for media with
    large losses, e. g., opague, strongly absorbing or
    (mie) scattering materials.

3. High costs and no packaged commercial equipment.

4. Tunability over the entire vibrational region and the
    observation of small Raman shifts are not routine
    operations.

5. There is always the possibility of sample damage with
    high-power lasers.

6. There are strong dependences of CARS signal or laser
    power, number density, and line width.

7. Interactions with neighboring resonances, background
    and electronic transitions may cause strong perturba-
    tion of the CARS spectrum.

REFERENCES

1. H. A. Szymanski, Ed., "Raman Spectroscopy," Plenum Press, New York (1967 and 1970), Vols. I and II.

2. A. Anderson, Ed., "The Raman Effect," Vol. 1, M. Dekker, Inc., New York (1971).

3. J. A. Konigstein, "Introduction to the Theory of the Raman Effect," (D. Reidel Publishing Co., Dordrecht, Holland, 1972).

4. G. Herzberg, "Molecular Spectra and Structure II - Infrared and Raman Spectra of Polyatomic Molecules," (D. Van Nostrand, Inc., Prince, N. J., 1945).

5. F. R. Dollish, W. G. Fateley, F. F. Bentley, "Characteristic Raman Frequencies of Organic Compounds," (John Wiley and Sons, New York, 1974).

6. D. S. Cain and A. B. Harvey, Dev. in Appl. Spectr. 7B, 94 (1970).

7. M. Lapp and C. M. Penney, Eds., "Laser Raman Gas Diagnostics," Plenum Press, New York (1974).

8. A. A. Boiarski, "Shock-Tube Diagnostics Utilizing Laser Raman Spectroscopy," NSWC/WOL/TR 75-53, White Oak Lab. (1975).

9. J. J. Barrett and A. B. Harvey, J. Opt. Soc. Am. 65, 392 (1975).

10. S. Lederman, M. H. Bloom, J. Bornstein, and P. K. Khosla, Int. J. Heat Mass Transfer 17, 1479 (1974).

11. M. E. Hillard, Jr., M. L. Emory, and A. R. Bandy, AIAA J. 11, 775 (1973).

12. M. Lapp, C. M. Penney, and J. A. Asher, "Application of Light-scattering Techniques for Measurements of Density, Temperature, and Velocity in Gasdynamics," ARL 73-0045, Wright-Patterson AFB, Ohio (1973).

13. R. W. Terhune, Bull. Amer. Phys. Soc. 8, 359 (1963).

14. P. D. Maker and R. W. Terhune, Phys. Rev. 137, A801 (1965).

15. J. J. Wynne, Phys. Rev. Lett. 29, 650 (1972).

16. F. DeMartini, G. P. Giuliani and E. Santamo, Optics Comm. 5, 126 (1972).

17. F. DeMartini, F. Simoni and E. Santamo, Optics Comm. 9, 176 (1973).

18. F. Capasso and F. DeMartini, Optics Comm. 9, 172 (1973).

19. R. W. Terhune and P. D. Maker, "Lasers," (A. K. Levin, Ed., Marcel Dekker, Inc., New York, 1968, Vol. II p. 295).

20. S. D. Kramer, F. G. Parsons, and N. Bloembergen, Phys. Rev. B, 9, 1853 (1974).

21. M. D. Levenson, IEEE J. Quantum Electron. QE-10, 110 (1974).

22. C. Flytzanis, "Theory of Optical Nonlinear Suscepti-bilities," Tech. Report 638, Harvard Univ. (1973).

23. M. D. Levenson and N. Bloembergen, J. Chem. Phys. 60, 1323 (1974).

24. M. D. Levenson and N. Bloembergen, Phys. Rev. B, 10, 4447 (1974).

25. M. D. Levenson, C. Flytzanis and N. Bloembergen, Phys. Rev. B, 6, 3962 (1972).

26. A. Nabara and K. Kubota, Jap. J. Appl. Phys. 6, 1105 (1967).

27. W. G. Rado, Appl. Phys. Lett. 11, 123 (1967).

28. J. Lukasik and J. Ducuing, Phys. Rev. Lett. 28, 1155 (1972).

29. F. DeMartini and J. Ducuing, Phys. Rev. Lett. 17, 167
     (1966).

30. J. J. Barrett and R. F. Begley, Appl. Phys. Lett. 27,
     129 (1975).

31. R. F. Begley, A. B. Harvey and R. L. Byer, Appl.
     Phys. Lett. 25, 387 (1974).

32. R. F. Begley, A. B. Harvey; R. L. Byer and B. S.
     Hudson, J. Chem. Phys. 61, 2466 (1974).

33. R. F. Begley, A. B. Harvey, R. L. Byer and B. S.
     Hudson, American Laboratory 6, 11 (1974).

34. S. E. Harris and B. S. Hudson, "(CARS) Coherent
     Anti-Stokes Raman Spectroscopy," Chromatix Applica-
     tion Note 6 (1975).

35. "Coherent Anti-Stokes Raman Spectroscopy (CARS),"
     Molectron Corp., Applications Note III.

36. B. Hudson, J. Chem. Phys. 61, 5461 (1974).

37. I. Itzkan and D. A. Leonard, Appl. Phys. Lett. 26,
     106 (1975).

38. J. E. Moore and L. M. Fraas, Anal. Chem. 45, 2009
     (1973).

39. F. Moya, S. A. J. Duret and J. P. E. Taran, Optics
     Comm. 13, 169 (1975)

40. P. Regnier, "Application of Coherent Anti-Stokes
     Raman Scattering to Gas Concentration Measurements
     and to Flow Visualization," Office National D'Etudes
     Et De Recherches Aerospatiales, 215 (1973).

41. P. R. Regnier and J. P. E. Taran, Appl. Phys. Lett.
     23, 240 (1973).

42. P. R. Regnier, F. Moya and J. P. E. Taran, AIAA J.
     12, 826 (1974).

43. P. R. Regnier and J. P. E. Taran, "Laser Raman Gas
     Diagnostics," edited by M. Lapp and C. M. Penney
     (Plenum Press, Inc., New York, 1974, p. 87).

44. P. Lallemand, "The Raman Effect," (A. Anderson, Ed.,
     Marcel Dekker, Inc., New York, 1971, Vol. I, p. 287).

45. D. L. Jeanmaire, M. R. Suchanski and R. P. Van Duyne,
     J. Am. Chem. Soc. $\underline{97}$, 1699 (1975) and subsequent
     articles in this series.

46. W. Kieter and H. J. Bernstein, Mol. Phys. $\underline{23}$, 835
     (1972).

47. W. Holzer, W. F. Murphy and H. J. Bernstein, J. Chem.
     Phys. $\underline{52}$, 399 (1970).

48. W. J. Jones and B. P. Stoicheff, Phys. Rev. Lett. $\underline{13}$,
     657 (1964).

49. P. Gadow, A. Lau, C. T. Thuy, H. J. Weigmann,
     W. Werncke, K. Lenz and M. Pfeiffer, Optics Comm. $\underline{4}$,
     226 (1971).

50. J. Klein, W. Werncke, A. Lau, G. Hunsalz and K. Lenz,
     Experimentelle Technik Der Physik $\underline{22}$, 565 (1974).

51. K. Kneipp, W. Werncke, H. E. Ponath, J. Klein, A. Lau
     and C. D. Thuy, Phys. Stat. Sol. $\underline{64}$, 589 (1974).

52. A. Lau, M. Pfeiffer, P. Gadow, W. Werncke, K. Lenz
     and H. J. Weigmann, Optics Comm. $\underline{4}$, 228 (1971).

53. C. D. Thuy, W. Werncke, A. Lau and K. Lenz, Experi-
     mentelle Technik Der Physik $\underline{22}$, 111 (1974).

54. W. Werncke, J. Klein, A. Lau, K. Lenz and G. Hunsalz,
     Optics Comm. $\underline{11}$, 159 (1974).

55. E. S. Yeung, J. Mol. Spectroscopy $\underline{53}$, 379 (1974).

56. J. F. Verdieck, S. H. Peterson, C. M. Savage and
     P. D. Maker, Chem. Phys. Lett. $\underline{7}$, 219 (1970).

57. D. Heiman, R. W. Hellwarth, D. M. Levenson and
     G. Martin, Phys. Rev. Lett., in press.

58. R. DeWitt, W. M. Tolles and A. B. Harvey, NRL Memor-
     andum Report, in preparation.

59. G. C. Bjorklund, IEEE J. Quantum Electron. <u>QE-11</u>, 287 (1975).

60. G. C. Bjorklund, Bell Lab., TM 74-1313-23 (1974).

61. J. W. Nibler, et al., unpublished results.

62. I. Chabay, G. K. Klauminzer and B. S. Hudson, Appl. Phys. Lett. <u>28</u>, 27 (1976).

63. B. S. Hudson, W. Heatherington, S. Cramer, I. Chabay and G. Klauminzer, Proc. Nat'l. Acad. of Sciences, to be published.

64. R. Wallenstein and T. W. Hansch, Optics Comm. <u>14</u>, 353 (1975).

65. R. T. Lynch, Jr., S. D. Kramer, H. Lotem and N. Bloembergen, "Double Resonance Interference in Third-Order Light Mixing," Optics Comm., in press.

PROGRESS IN THE VIBRATIONAL SPECTROSCOPY OF

COMPLEX SYSTEMS

Bernard J. Bulkin, Dolores Grunbaum,
Arthur Noguerola, and Nehama Yellin

Polytechnic Institute of New York
333 Jay Street, Brooklyn, N.Y. 11201

and
Hunter College
695 Park Avenue
New York, N.Y. 10021

In the last decade, there has been a revolution in the practice of vibrational spectroscopy. This has come about in the practice of both infrared and Raman spectroscopy. The Raman revolution, began with the introduction of continuous gas lasers, improvements in photomultipliers and double monochromators, has continued with the introduction of holographic gratings and the coupling of instrumentation to minicomputers. The infrared revolution is tied to the minicomputer as well, particularly for the practice of Fourier transform infrared spectroscopy.

With the increased capability of infrared and Raman instrumentation have come a wide variety of theoretical and experimental developments in vibrational analysis. In this paper we discuss a few of these which have been of concern in our laboratory.

Raman Polarization Measurements: It has long been recognized that Raman scattering contains

structural and chemical information of interest
in the depolarization ratios of Raman bands.
After the introduction of laser excited Raman
techniques, this was exploited in the assignment
of lattice modes of crystals. More recently,
we have been making extensive use of depolarization ratios obtained on our automated Raman instrument (1) for structural studies.

The principle behind most of our structural
studies is a simple one. If a molecule has no
symmetry elements other then the identity, (C_1
point group) then all vibrational modes belong to
the A representation, are Raman active, and are
polarized. If, on the other hand, the molecule
has a plane of symmetry as the only symmetry
element, its vibrations transform according to
the C_s point group. In this case, some vibrations will have A' symmetry, while others will
have A" symmetry. All modes will still be Raman
active, but A' modes will be polarized while A"
modes will be depolarized. A third interesting
use in low symmetry molecules is C_i symmetry, in
which a center of inversion is the only symmetry
element. In this case modes can belong to A_g or
A_u symmetry. Only A_g modes are Raman active,
and all are polarized.

The problem with applying Raman polarization
selection rules to complex, low symmetry molecules
is that the number of vibrational modes is usually
large, frequently with overlapping bands. It is
necessary to determine the ratios accurately for
all bands. This is most easily done in a format
first discussed by Scherer (2), in which the
spectrum usually referred to as $I_\perp$ is appropri-
ately scaled and subtracted from the $I_\parallel$ spectrum
to produce a display containing only the polarized
bands. One can then visually compare this display
with that of $I_\parallel$ to check if all bands are present
in both.

We have demonstrated the utility of this with
several examples. Figure 1 shows as an example
the Raman spectrum of dimethyl sulfoxide in the

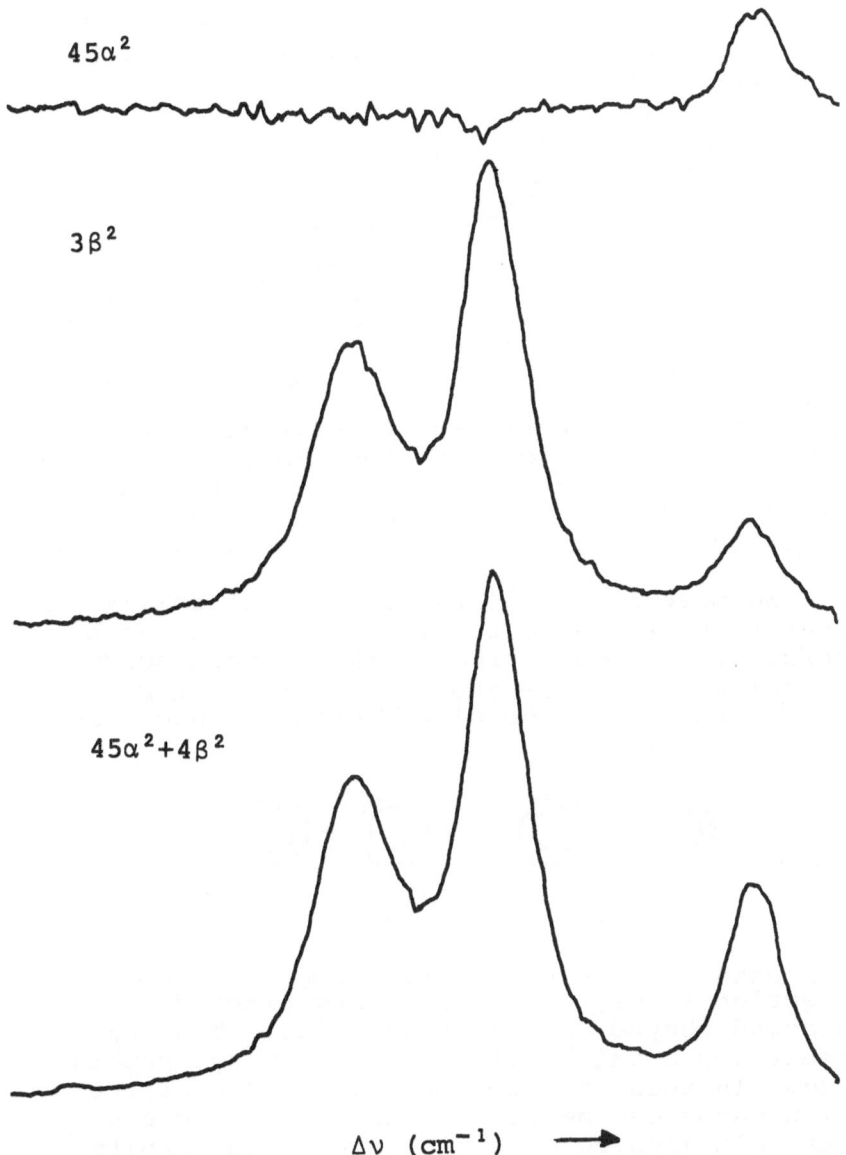

Figure 1. Raman spectrum of dimethyl sulfoxide,
200-400 cm^{-1} region.

$200-400 \text{cm}^{-1}$ region. This shows clearly the presence of one polarized and two depolarized bands. In work published elsewhere (3), we examined the spectrum of 4,4'-bis-methoxy-azoxybenzene,

which had been determined by X-ray diffraction to have non-coplanar benzene rings in the crystalline phase. A Raman solution spectrum containing more than 100 bands showed all bands to be polarized, consistent with the non-planar structure

We have also utilized the polarizability separation procedure to study several cases of conformation about carbon-carbon single bonds. Molecules such as 2- and 3-halobiphenyls as well as the methyl homologs were examined. In

each case, the 1-substituted compound obeyed C_1 selection rules, while the 3-substituted compound obeyed C_s selection rules. The same result was obtained with 1 and 2 phenyl naphthalene. In these spectra more than 50 distinct Raman bands can be seen. Finally, ortho and meta tolualdehyde were examined. The results indicated that in the meta compound C_s selection rules are obeyed, indicating that the -C=O (H) moiety lies in the plane of the ring, whereas in the ortho compound it does not.

The separation of isotropic and anistropic polarizability is useful for two other cases. Spiro and Strekas (4) demonstrated the presence of inversely ($\rho > 3/4$) polarized bands in the resonance Raman spectra of cytochrome c and hemoglobin. The polarization separation display is quite convenient for finding such modes, which appear as negative deviations from the baseline in the "$45\alpha^2$" display. Second, it should be noted that the Rayleigh scattering occurring close to the laser exciting line is generally depolarized, and low frequency polarized modes, hidden by the exciting line, are sometimes revealed in the $45\alpha^2$ spectrum.

Raman Spectra of Biological Membranes: We have previously described the Raman spectroscopic changes which occur upon heating of phospholipid-water gels, a multilayer system with some importance to the study of biological membranes (5). These changes are related to hydrocarbon chain melting. In our earlier work, the uncertainty in relative intensities was greater than 10%. While this was acceptable for understanding gross changes in the spectra, more accurate data were needed to examine these changes in more detail. Through signal averaging one can accomplish this. We have now been able to accumulate sufficient signal so that relative intensities are accurate to within 1%. As a result, phase transitions were detected which were not previously seen. In these transitions the overall intensity changes range from 4 to 20%. These phases are seen in thermodynamic measurements, and we can now see how they manifest themselves in the Raman spectrum as well. Details will be published elsewhere (6).

In carrying out such experiments by signal averaging, we use a low laser excitation power. This is desirable, as small heating effects must be eliminated in the study of phase transitions. By using the computer to signal average we can achieve the dual aims of low laser power and high intensity accuracy. In general, Raman spectrometers

are sufficiently stable that multiple scanning is
unnecessary. Instead, one incremental scan at tne
desired integration time can be made.

These systems also represent an interesting
case of Raman activity which is related to the
depolarization ratio measurements discussed
earlier. In the lower temperature phase, the
hydrocarbon chains are in all trans, extended
conformation. The spectrum of such a phase in
the C-H stretching region is shown in Figure 2.
Here the strongest oands are undoubtedly due to
symmetric (2844 cm^{-1}) and asymmetric (2877 cm^{-1})
CH_2 stretching vibrations, and some of the
intensity near 2960 cm^{-1} arises from the CH_3
stretches. Even in the all trans conformation,

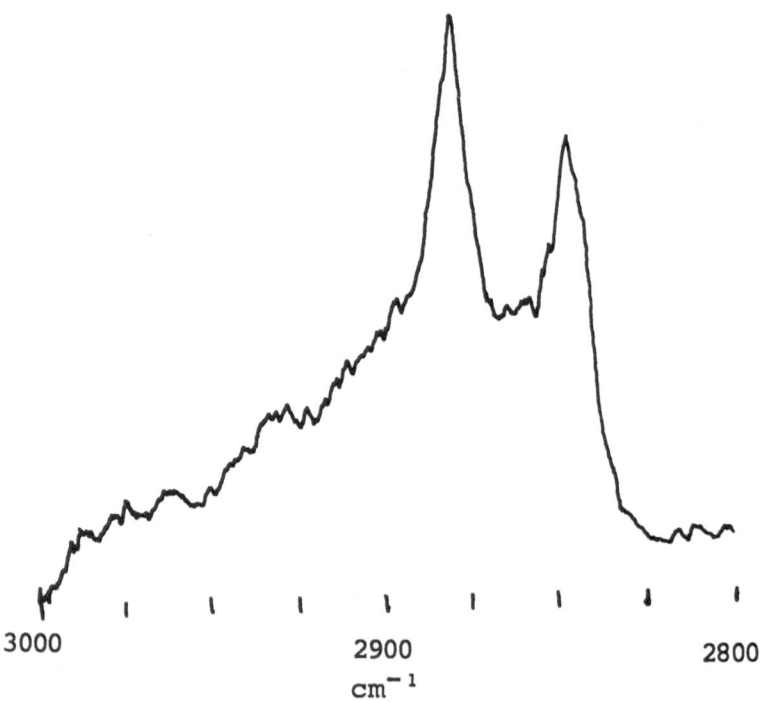

Figure 2.   Raman spectrum of DPL: 31% H_2O gel
at 24° C.

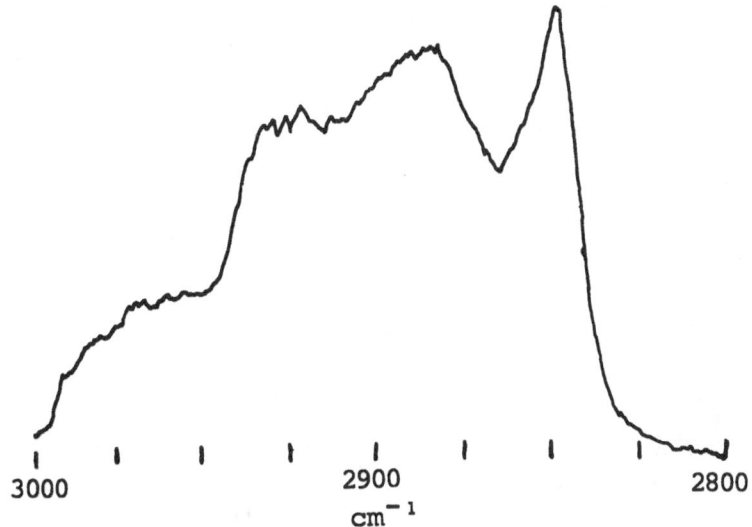

3000   2900   2800
                                  cm⁻¹

Figure 3.   Raman spectrum of DPL: 31% liquid
crystal at 70° C.

Even in the all trans conformation, the coupling
between CH_2 groups is weak for CH stretches, ex-
tending over no more than 2-4 carbon atoms.

In Figure 2, weak scattering can be seen in
the 2930 cm⁻¹ region. When the temperature is
raised above the phase transition this becomes
much more intense, as can be seen if Figure 3.
Other changes also take place, such as small
frequency shifts and intensity increases at
other frequecies. While the explanation of some
of these is complex due to the large number of
overlapping bands, the increase in intensity
at 2930 cm⁻¹ is definitely associated with a
lowering of symmetry of the CH_2 groups, as this
band is quite intense in the infrared spectrum
of the lower temperature phase.

To show the spectroscopic changes which occur
in the 2800-3000 cm⁻¹ region with temperature
most clearly, we have, in Figure 4, scaled the
high temperature spectrum so that the intensities

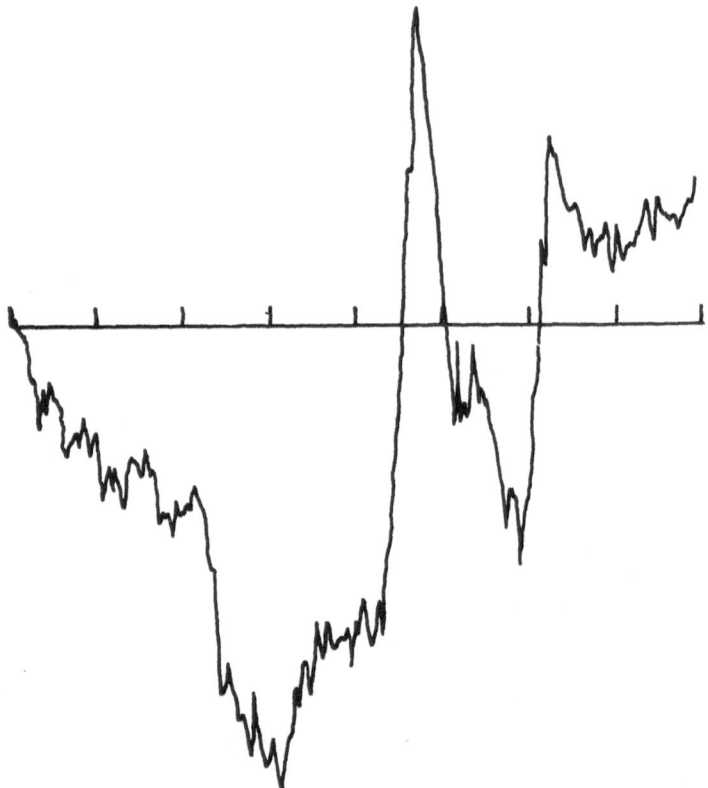

Figure 4.   Difference spectrum, 2800-3000 cm^{-1}
region, for DPL-H$_2$O mixture.   Figure 3 spectrum
has been scaled using the lowest frequency
maximum, and subtracted from spectrum of Figure 2.

at the "2844 cm^{-1}" maxima are equal, and subtracted the spectra.   At the lower frequency and,
the pattern which looks like a first derivative
is a manifestation of the frequency shift of a
few wave numbers which this band undergoes at
the phase transition.   At the high frequency end,
particularly at 2900 and 2930 cm^{-1}, we see the
great increase in intensity referred to above.

Lattice Vibrations of Organic Crystals: There
have been a number of attempts to compute lattice
vibration frequencies. Without reviewing these
here in any detail, one can say that they may be
divided into two classes. In one, a limited
number of force constants are permitted to take
on nonzero values. These are adjusted to give
the best fit to the observed frequencies. Usually,
this has been carried out in cases where only
infrared or Raman data are available, and the
normal modes are factored by symmetry. In the
second approach, a potential function has been
determined, often semiempirically, using the
physical properties of the crystal such as heat
of sublimation and equilibrium structure. In
principle, this treatment should give more information about the intermolecular forces responsible for the lattice vibrations. It is considered
to be a more laborious approach, as one must
derive analytical expressions for the force
constants by hand. In addition, there have been
some problems with the theory, or again, with
the lack of sufficient experimental information,
or both. Two comprehensive reviews, with
extensive references, have been published.

We have been pursuing calculations using the
second method, with the viewpoint that atom-atom
interation potentials now have been sufficiently
well parameterized that they can be used without
iterations involving the experimental frequencies
or structures. In this sense the calculations
are an experimental tool much like parameterized
semi-empirical molecular orbital calculations.

In our work the potential functions of the
form

$$V_n = \frac{-A}{r^6} + Be^{-Cr}$$

have proved most useful. We have summarized
elsewhere (7) where parameters such as A, B,
and C for the n th atom-atom interaction may

be found in the literature. Other authors have
used potentials of the closely related form

$$V_n = \frac{-A}{r^6} + \frac{B}{r^D}$$

as well as static dipole-dipole terms and coulom-
bic potentials where appropriate. Such calculations can, in any case, now be done with minimal
computer programming - to yield interesting
results on the lattice dynamics in organic
crystals.

Acknowledgements: The research described herein
was supported by grants from the U.S. Army
Research Office and the American Cancer Society.

REFERENCES

1.   B. J. Bulkin, E. Cole, A. Noguerola, J. Chem.
     Ed., $\underline{51}$, A273 (1974)

2.   J. R. Scherer, S. Kint, G. F. Bailey, J. Mol.
     Spectr., $\underline{39}$, 146 (1971)

3.   B. J. Bulkin, D. L. Beveridge, F. T. Prochaska,
     J. Chem. Phys., $\underline{55}$, 5828 (1971)

4.   T. G. Spiro, T. C. Strekas, Proc. Nat. Acad.
     Sci., USA, $\underline{69}$, 2622 (1972)

5.   B. J. Bulkin, N. Krishnamachari, J. Amer.
     Chem. Soc., $\underline{94}$, 1100 (1972)

6.   B. J. Bulkin, N. Yellin (to be published)

7.   B. J. Bulkin, D. Grunbaum, J. Phys. Chem.,
     $\underline{79}$, 821 (1975)

MEASUREMENT AND APPLICATIONS OF FLUORESCENCE LIFETIMES IN THE NANOSECOND RANGE USING THE TIME CORRELATED SINGLE PHOTON TECHNIQUE

L. A. Shaver and L. J. Cline Love

Department of Chemistry, Seton Hall University
South Orange, New Jersey 07079

The time correlated single photon (TCSP) technique is more commonly known as the "single photon counting" method, a term that has led to confusion with conventional photon counting detection. Both are spectroscopic detection methods but they differ greatly in principle, instrumentation, and application. Perhaps the distinction between the two techniques will be served by first briefly describing photon counting. Photon counting is an instrumental technique for measuring low levels of light or particle beam densities (1-5). Individual photons incident on a photomultiplier tube (PMT) create discrete pulses or bursts of electrons arriving at the anode at a rate directly proportional to the photon flux.   The instrumentation is relatively straightforward. The current pulse from the PMT anode is converted to a voltage, amplified, and sent to a pulse height discriminator. Only those pulses within preset voltage limits are passed on to a counter. Advantages of photon counting include: 1) the signal-to-noise (S/N) ratio is enhanced compared to the DC current measurement method, 2) the data are in digital form, 3) the sensitivity is greater at low light levels, 4) the measurements may be integrated over long periods of time resulting in higher precision and accuracy (2).

Photon counting has been used for low-level light measurements in Raman (6), molecular and atomic fluorescence (7,8), absorption spectroscopy (1,2), and astronomy (9). Other quantum detectors such as electron multipliers and particle detectors may also be used with photon counting systems. The technique has been used in mass spectrometry, X-ray measurements, nuclear particle analysis, laser research and molecular beam spectroscopy (4,5).

The time correlated single photon technique, which is the subject of this paper, is an instrumental method for measuring emissive intensity decay in the nanosecond time range. The technique has been reviewed in the literature (10-16). Figure 1 is a schematic diagram of a time correlated single photon instrument designed to measure fluorescence decay curves. The principle of operation of the TCSP is illustrated in this example. Two photo-multiplier tubes are used, one to observe the excitation source directly and the other to view the sample. Upon the initial rise in intensity of the excitation source the first PMT sends a pulse to a time-to-amplitude converter (TAC) which triggers the voltage ramp shown in Figure 2. Meanwhile, excited molecules in the sample begin to decay and luminescent emission is observed by the second PMT. A single luminescent photon incident on the second PMT cathode creates a pulse which is sent to the TAC. This triggers the TAC to halt the voltage ramp and to output a pulse with a voltage directly proportional to the elapsed time between trigger pulses. The TAC output pulses are sent to a multichannel pulse height analyzer (MCA) where they are sorted according to amplitude and stored as counts in the memory. Because the amplitudes of the TAC output pulses are directly proportional to elapsed time, the channels of the MCA correspond to increments of time. The process of excitation is carried out several thousand times per second. Following each excitation cycle the TAC is reset and the process is continued until sufficient counts have been collected in the MCA to define the decay curve. The functions of the discriminator, amplifier, and delay calibrator modules shown in Figure 1 are discussed in a later section. It is important to note that the lumines-cence is attenuated so that no more than one photon per excitation pulse reaches the PMT photocathode. The theoretical basis for this requirement will be detailed when probability considerations are discussed.

The decay curve of emission intensity versus time is built up over a period ranging from several minutes to a few hours in a time correlated manner. The time at which the emission intensity is greatest corresponds to the time of highest probability of detecting an emission photon by TCSP technique. The counts accumulated in the MCA are the

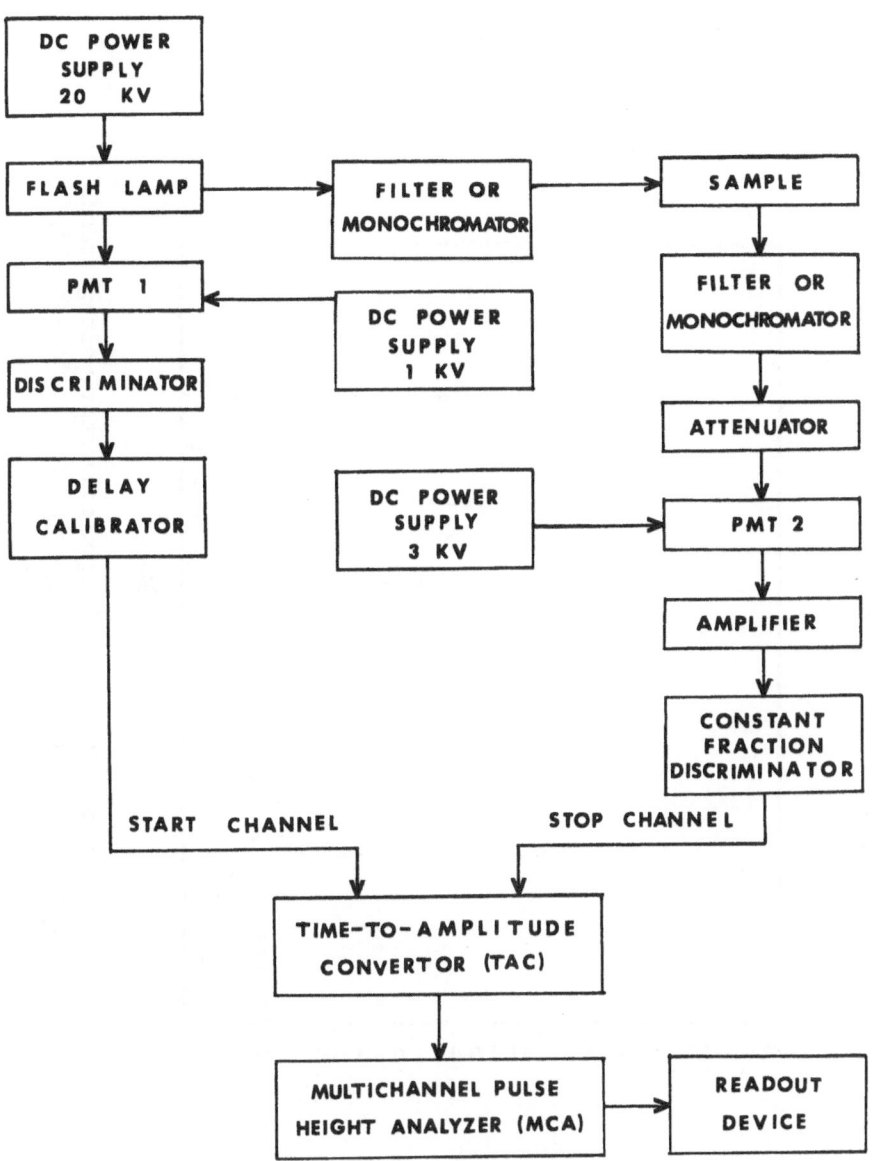

Figure 1.   Block diagram of a time correlated single photon nanosecond fluorometer.

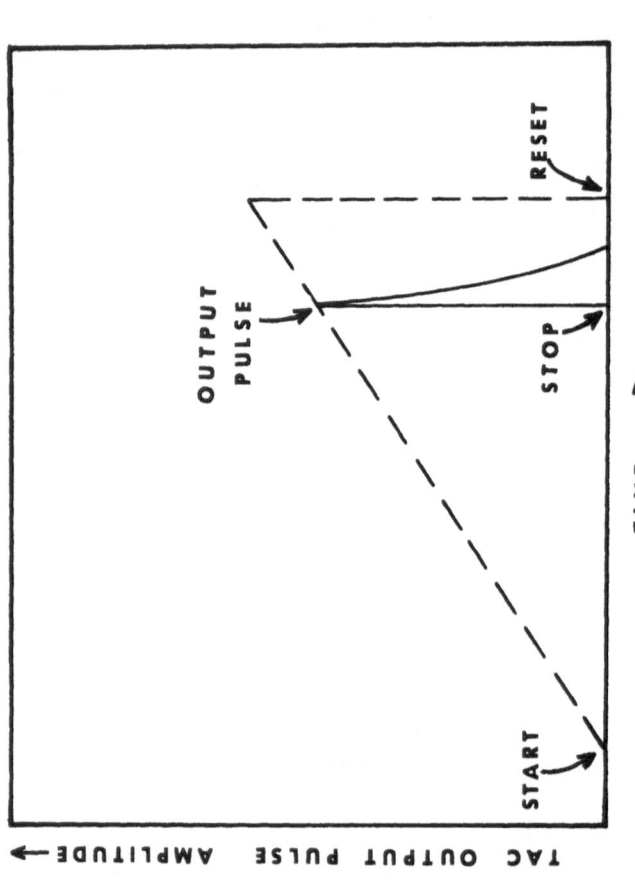

Figure 2. Operation of time-to-amplitude convertor voltage ramp.
(Courtesy of Analytical Chemistry from reference 15)

relative emission intensities corresponding to the proba-
bilities of luminescence on the nanosecond time scale.

Three other techniques which have been used to measure
luminescence decay deserve mention. The simplest technique
uses a PMT, a fast oscilloscope, and either a camera or
recorder to detect the decay curves (17,18). Another
technique uses a modified image converter tube to measure
luminescence intensity as a function of time (16). A
third technique uses a modulated excitation source to
permit measurement of the phase shift of the luminescence
(10). The phase shift method does not result in a decay
curve but the information can be used to calculate the
luminescent lifetime. The time correlated single photon
technique has several advantages over these methods. The
TCSP method has time resolution of about 0.2 ns and the
S/N ratio is improved by signal averaging. Also the technique usually has adequate sensitivity due to the small
number of photons required for detection and any time
jitter in the pulsed source is averaged over many cycles.
Although other techniques may be superior in one or two
ways, the TCSP method is least limited in overall applicability.

The method was developed in 1961 by Bollinger and
Thomas (19) to measure the lifetimes of scintillators
excited by different types of high energy radiation. In
the same year Koechlin (20,21) extended the resolution of
the method into the nanosecond region. Since then improvements have been made in pulsed source technology, photo-
multiplier tubes and associated electronics. State-of-
the-art modules may be purchased commercially and the
technique implemented by skilled laboratory personnel.
The principal use of TCSP technique among chemists is for
obtaining the luminescent lifetimes of molecules. The
lifetime is usually defined as the time needed for the
luminescence light intensity to decrease by the fraction
1/e, where e is the base from the Napierian logarithm
system. Often this information must be extracted from
complex experimental decay curves by digital computer
calculations. Several computer algorithms have been
developed for this purpose which require expert computer
programming capability and careful judgement to interpret
the results. Such calculations are commonly referred to
as deconvolution and are discussed in a later section.

The measurement of fast photophysical and photochemical phenomena can provide information regarding the
nature of intra- and intermolecular and atomic interactions.
Other experimental parameters such as quantum yield, Φ,
may be used in conjunction with the lifetime, τ, to
mathematically derive energy considerations concerning the
excited state. This information is helpful in understanding the mechanism of excitation and deactivation processes.
A molecule or atom which has been activated to some excited
state M* by absorption of radiant energy will remain in
this high energy state for variable lengths of time.
Deactivation can proceed by numerous mechanisms such as
chemical reaction, radiationless transitions, fluorescence,
and phosphorescence. Fluorescence generally occurs within
10^{-6} to 10^{-12} seconds of activation. Conventional fluorometers are capable of measuring decay characteristics of
fluorophors in the microsecond range. The TCSP arc-dis-
charge flashlamps, has extended fluorometry into the nanosecond range and lifetimes of approximately 10^{-10} seconds
can be measured reliably.

THEORY

Statistical Considerations

In the time correlated single photon technique it is
essential that no more than one emission photon per
excitation pulse be detected. The consequence of multiple
photon events is the so-called "pulse pileup". The MCA
counts accumulate disproportionately in the early portion
of the decay curve, resulting in badly distorted data.
The reason pileup occurs is found in the nature of the
time-to-amplitude convertor operation. The TAC voltage
ramp is initiated by the start signal and is halted when
the stop signal is received. The TAC then recycles after
a preset time interval and then waits for another start
trigger. A later second emission photon from the same
excitation pulse arriving at the stop PMT will not be
counted because the TAC is disabled. Thus the decay curve
is biased toward early arriving photons. The emission
beam must be attenuated such that probability of detecting
more than one photon per excitation pulse is small. Thus,
the detection of a photon is a rare event and the probability is calculated from Poisson's equation

$$P_n = \frac{m^n}{e^m \, n!} \tag{1}$$

where P_n is the probability of detecting n photons per excitation pulse and m is the average number of photons detected per excitation pulse. Multiple photon pileup error may best be appreciated by examining the Poisson probability polygons shown in Figure 3. Note that when the average counting rate is one photon per excitation pulse the probability of detecting two photons is 0.184, which is large enough to cause severe pileup. In practice an average count rate of 0.01 to 0.05 photons per excitation pulse is used and the probability of multiple photon events is very small. Coates (22), Davis and King (23), and Donohue and Stern (24) have reported methods for the mathematical correction of pileup. If an average count rate of 0.05 is maintained then there is little need for pileup correction. This small average count rate is a major disadvantage of the TCSP technique. Ordinarily, between 10^3 and 10^6 counts are accumulated in the peak channel. If the excitation source pulse repetition rate is of the order of a few thousand pulses per second, it can take hours to obtain a single decay curve. Knight and Selinger (14) discuss the use of instrumental pileup detector modules which work by rejecting and TAC output for a detected emission photon arrival if a second photon arrives within the TAC sweep range. The use of pileup detectors enables increased average count rates to be used without distortion of the decay curve. However, as the average count rate is increased to greater than one photon per cycle, multiple photon events increase, more TAC outputs are rejected, and data accumulation time is actually greater than for lower average count rates.

An alternate way of decreasing data collection time is increasing the excitation source repetition rate. Zimmerman and coworkers (25) report the use of a flashlamp (width at half maximum 3.5 ns) operating at 20 kHz and at an average count rate of 0.05, which required only 5 min. to collect 2000 counts in the peak channel of the decay curve for relatively short-lived fluorophors.

Various properties of the sample, such as stability and fluorescence quantum yield, must also be considered

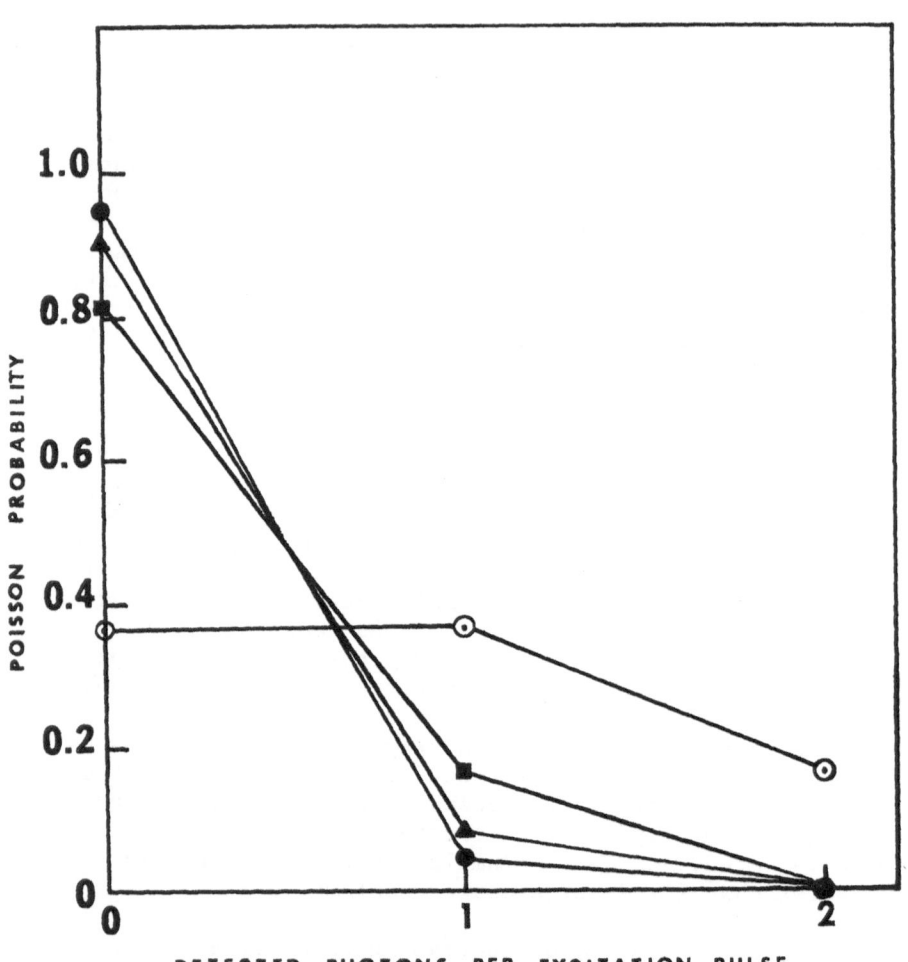

Figure 3.   Poisson probability polygons for various
average counts per excitation pulse (◐)
1.0, (■) 0.2,(▲) 0.1, and (●) 0.05
(Courtesy of <u>Analytical Chemistry</u> from
reference 15)

when optimizing the count rate. If the emissive properties
of the sample change with time then an abbreviated experimental period is required. If this is the case, a 0.20
count rate could be used to rapidly accumulate data, which
must then be mathematically corrected for pulse pileup.
Fluorophors that show strong emission may be analyzed
after appropriate sample dilution to obtain a 0.05 count
rate.

   Noise. The decay curve that is built up in the MCA
is based on the probability of detecting emission photons
along the time axis. The greater the emission photon flux,
the greater the probability of photon detection by the
TCSP instrument. Ordinarily about 100 to 500 channels of
the MCA are used and data collection continues until a
relatively smooth decay curve is obtained. Most MCA's
have oscilloscope screens that display memory contents.
However, decay curve data often must meet more stringent
requirements than just being pleasing to the eye. The
relative intensity of the decay curve varies over 3 or 4
orders of magnitude from less than 100 counts at each
extreme to between 10^3 and 10^6 counts in the peak channel.
For data collected by a TCSP instrument, the count total
in each channel is an estimate of the mean of a Poisson
distribution of counts for that channel (26). Also, the
variance of a Poisson mean is equal to the number of counts
in the channel, hence the standard deviation equals the
square root of the number of counts in that channel. Such
noise is random and normally distributed.

   The statistical noise distribution about a hypothetical decay curve is shown in Figure 4. The limits are 3
standard deviation units which include approximately 99%
of the data. There is an advantage to accumulating a large
number of counts, due to the resultant improvement in S/N
ratio. However, increased count accumulation is obtained
at the cost of experimental time. Depending on experimental conditions an additional decade of peak channel
counts often may take an hour or more additional time to
collect. Obviously a trade off between S/N ratio and
experimental time is required.

   The minimum S/N ratio which is acceptable must be
clearly defined if the data are to be deconvoluted by one
of the several computer alogorithms available. If decon-

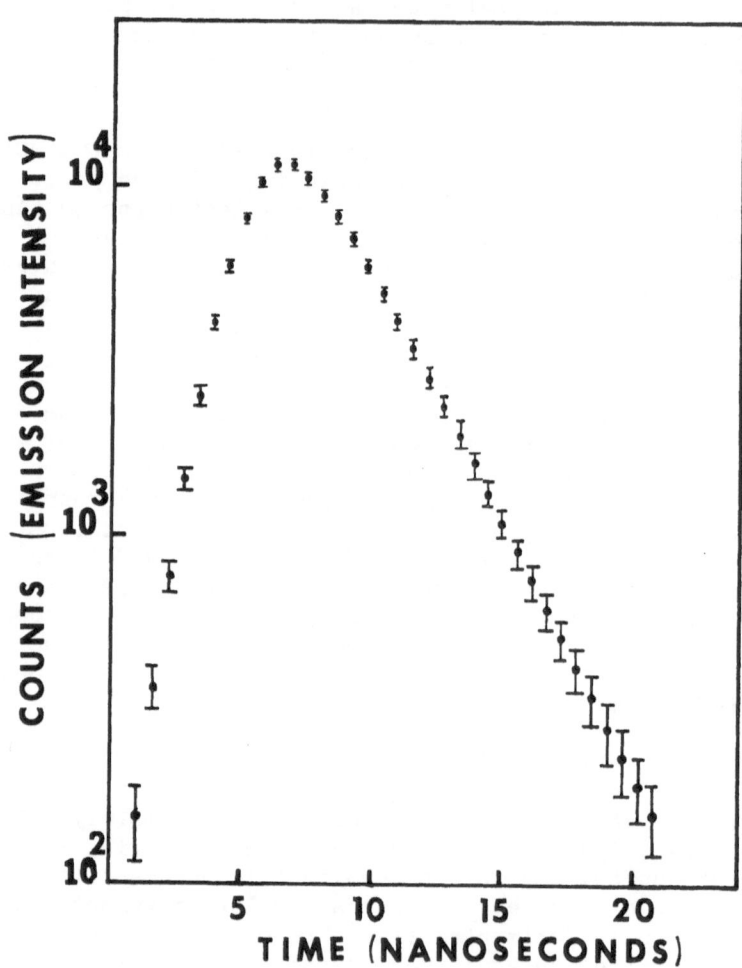

Figure 4.   Statistical noise distribution on a simulated
            decay curve.   The limits shown are three
            standard deviation units.

volution is used to resolve mixtures of fluorophors, to
calculate lifetimes of flurophors less than the source
lifetime, or to calculate nonexponential decays, then S/N
ratio requirements are more stringent than for cases of
relatively long-lived single exponential decays which need
only a graphical slope evaluation of a semilogarithmic plot
of the data. Decay curves having different number of counts
in the peak channel can be simulated by digital computer
programs. If noise is superimposed on these simulations
and the curves subsequently deconvoluted, information about
the experimental data requirements can be ascertained.
Simulations in our laboratory (27) show that about 10^4
counts in the peak channel are necessary for reliable
deconvolution of two overlapping exponential decay curves
having similar lifetimes.

INSTRUMENTATION

The instrumentation for TCSP measurements is commercially available (13). The necessary equipment may be
conveniently assembled from appropriate modules, most of
which can be used for other experimental purposes as well.
Some workers find it convenient or occasionally mandatory
to construct portions of the apparatus to meet experimental
or cost requirements. This section deals with the principal
components of a nanosecond fluorometer used to measure
fluorescence lifetimes. With slightly different instrumental configurations, the same principles of the TCSP
technique may be used in the measurement of other time
correlated phenomena such as scintillator decay characteristics (19).

The elements common to all TCSP instruments are the
start and stop circuits and the time-to-amplitude convertor.
In the diagram of a nanosecond fluorometer assembly shown
in Figure 1, each block represents a separate module.
Some modules such as PMT power supplies, monochromators,
and filters are conventional and will not be discussed.
Detailed attention will be given only to the specialized
equipment shown in the diagram.

Most nanosecond fluorometers use spark discharge
flashlamps operating at a few kHz to excite singlets in
the sample. Two PMT's are used, one observing the flash-

lamp, and the other measuring the emission. PMT signals
are sent to discriminators, and amplifiers and then to the
TAC. The TAC signal is sent to a multichannel analyzer
for storage, and the MCA contents are eventually transferred to hard copy via a readout device such as teletype-
writer or tape punch.

Excitation Sources

Nanosecond Flashlamps. Typical flashlamps used for
nanosecond fluorometry consist of two tungsten electrodes
held a few mm apart. High voltage is applied across the
electrodes and a spark discharge occurs at the breakdown
potential. Typically, lamps operate between 1 and 30 kHz
and have a full width at half maximum (FWHM) between 2
and 4 ns. This repetitive flashing may be obtained in one
of two ways. One way is to operate the lamp in a free-
running relaxation circuit. Yguerabide (16) describes in
detail the operation of free-running lamps, and the basic
circuit is shown in Figure 5a. The flashlamp has an
inherent capacitance usually in the picofarad range. High
voltage on the order of a few kV is applied and the flashlamp capacitor is charged through the current limiting
resistor, R. When the breakdown voltage is reached the
upper electrode discharges, producing an intense light
pulse. The frequency of the lamp depends on the applied
voltage, gap length, fill gas type and pressure, and the
RC time constant. The temporal and spectral characteristics of the discharge are determined by excited-state
decay of the fill gas. The intensity of the discharge
increases with increases in gap length, fill gas pressure,
and capacitance (16). The repetition rate is usually controlled by changing the applied voltage. The total output
is approximately 10^7 photons per pulse (13).

The second method of operating a nanosecond flashlamp
is by gating the discharge. Ware (11,12) discusses the
use of thyratrons and the circuit is shown in Figure 5b.
The thyratron does not conduct until a gating pulse is
applied to its grid. The conductance of the thyratron is
then suddenly increased and the lamp discharges rapidly.
The charging, gating, and discharging is repeated at a
frequency controlled by the thyratron gating pulses. The
circuitry for the gated lamp is quite complex compared to

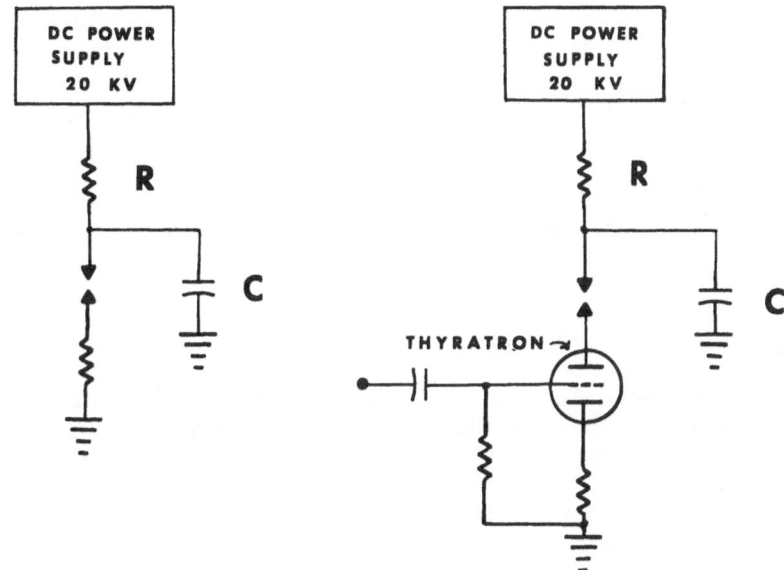

Figure 5. Schematic diagram of (a) free-running
flashlamp circuit and (b) gated flashlamp circuit.

the free-running lamp. The advantages of gating are careful control over the repetition rate, and more intense
discharge due to greater charging voltages. Accurate
control of the frequency of the lamp is not critical for
many applications of the TCSP measurement technique.

Fill gases used in flashlamps have included air, N_2,
D_2, H_2, Ar, and He. These gases are used to charge the
electrode chamber at pressures between 1 and 20 atmospheres.
The particular electrode material, fill gas and pressure
in the lamp will determine, to a large extent, the wavelength range, intensity and pulse durations of the flash.
The goals of improvements in flashlamp technology are to
decrease FWHM of the pulse to less than a nanosecond,
increase the repetition rate to 20 kHz or greater, and
provide a spectral output suitable for fluorescence
studies.   Details of the construction of flashlamps may
be found in the literature (14,16,28).

Pulsed Lasers. Pulsed lasers offer the two advantages of higher intensity and shorter duration pulses over
flashlamps. For nanosecond fluorometry with TCSP detec-
tions the higher intensity is of interest for extremely

weak fluorophors. The advantage of shorter pulse width is
particularly important for exciting subnanosecond-lived
fluorophors. The narrower the excitation pulse the shorter
luminescent τ's that can be measured. Four general types
of lasers that have been applied recently to nanosecond
luminescence studies are cavity dumped (29), pulsed (30),
Q-switched (31), and mode-locked (32-35). The use of
laser excitation sources should increase as the cost factor
decreases and available wavelength ranges increase.

Photomultiplier Tubes

The PMT used for the start channel views the flashlamp directly. The initial rise of the discharge creates
a burst of electrons in the PMT of sufficient current to
pass the threshold of the start channel discriminator.
This pulse is used to initiate the voltage ramp in the
TAC. A photomultiplier tube such as the RCA 1P28 can be
used for the start channel (11,36). Alternatively, the
start channel pulse can also be obtained from the terminat-
ing resistor of the flashlamp circuit (11-13,16). There
is no apparent advantage in using either method of obtaining the start pulse.

The stop channel PMT should be one specially designed
for TCSP measurements. Ideally, the tube should have a
high gain (10^6-10^8), have a fast rise time (~2 ns), low
dark current, and be of end window design. The Philips
56DUVP/03 (14) and RCA 8850 (13) are two suitable PMT's.
The PMT should be mounted in a special base/housing such
as the ORTEC 9201, containing an appropriate voltage
divider network (37) and a RF shielded housing. A
kilovolt variable power supply is required for most
commonly used PMT's. The time resolution of TCSP technique
depends partly on the proper adjustment of the focusing
voltage on the PMT (11).

Electronic Timing Circuitry

Many of the electronic circuitry modules are conventional pulse amplifiers and discriminators. Specific
timing devices like the TAC and nanosecond delays may be
obtained from manufacturers such as ORTEC, Inc. (13).

The start channel electronics consists of a discriminator and delay calibrator. The discriminator is a leading-edge type designed for high-frequency photomultiplier current inputs. The output of the discriminator is a single timing pulse of constant amplitude and width. The ORTEC Model 436 100 MHz Discriminator is commonly used and is suitable for either PMT or electrically derived pulses. The delay calibrator module is a series of variable nanosecond delay lines that synchronize the timing of start and stop channel pulses. The ORTEC Model 425 Delay is variable between 0 and 31 ns. The unit consists of coils of RG58/U cable between 8 and 128 inches in length.

The stop channel electronics consists of an amplifier and discriminator. In this channel a single photon pulse from the PMT is used for timing. This pulse is amplified by a unit such as the ORTEC Model 454 Timing Filter Amplifier. The discriminator is the ORTEC Model 463 Constant Fraction Discriminator. Constant fraction discrimination (CFD) is important in the TCSP technique in eliminating "time jitter" and improving resolution. Knight and Selinger (14), and Ware (12) compare CFD to leading edge discrimination (LED) and explain the importance of CFD in timing applications. The pulses from the stop channel PMT will have different energies because photons incident on different portions of the photocathode produce different pulse amplification (14). Pulses having different amplitudes will cross the lower level of an LED at different times creating timing jitter. In CFD the threshold is reached when the pulse has reached a fixed fraction of its maximum. Thus, the discriminator operates to give timing pulses at exactly the same time during each flashlamp cycle for photons arriving at the photocathode at exactly the same time, regardless of the respective pulse amplitudes.

The heart of the electronics is the time-to-amplitude convertor. The operation of the TAC is shown in Figure 2. When the "start" pulse is received the TAC initiates the voltage ramp. When a "stop" pulse is received the ramp is halted and the TAC outputs a pulse having a voltage directly proportional to the elapsed time between "start" and "stop" signals. This pulse is sent to a multichannel pulse height analyzer and stored as one count in the channel corresponding to that pulse height. If no stop pulse is received during the sweep cycle the TAC recycles to wait for another start pulse. The ORTEC Model 457 Biased Time to Pulse Height Convertor is equipped with a bias amplifier that linearly increases the output voltage so that the decay curve can be stored in any selected series of MCA channels.

Most MCA's are compatible with TCSP equipment. The unit should be capable of storing at least 10^4 counts in each channel, have at least 100 channels, be capable of transferring data in digital form to a readout device such as a typewriter, paper tape punch, or magnetic tape. It should have an oscilloscope display to monitor memory contents and to allow rapid photographic recording of the decay curves. Some units are capable of driving X-Y recorders to produce hard copy displays of the memory contents.

Zimmerman and coworkers (36) used a PDP/8 computer instead of a MCA. The TAC signal was digitized by an A/D convertor then stored in the computer memory. The use of the computer has the advantage that deconvolution can be performed directly without transferring the data.

Instrumental Performance

The resolution of the TCSP technique is limited by the uncertainty in the transit time between the electron emission at the stop channel PMT photocathode and the arrival of the electron pulse at the anode. Birks and Munro (10) estimate the uncertainty at about 0.3 ns. This uncertainty can be reduced to 0.2 ns by masking all but the central portion of the photocathode (14). Yguerabide (16) shows the experimental resolution depends on the TAC time and output voltage ranges, on the MCA voltage scales and the number of MCA channels used. Time jitter also results in loss of resolution. The ORTEC Model 457 TAC operates at full-scale ranges from 50 ns to 80 µs.

The overall precision of the TCSP method for lifetimes in the 10^{-8} and 10^{-6} s range is 2 percent (14). The precision at the 10^{-10} and 10^{-9} s levels are about 50 percent and 5 percent respectively (36). The accuracy of measuring lifetimes decreases with decreasing lifetime.

For lifetimes that are less than or nearly equal to the
decay lifetime of the flashlamp the accuracy depends on
the deconvolution method applied. Computer simulations
show that lifetimes down to 50 ps can be deconvoluted with
10 percent error for a flashlamp with a 1.1 ns decay lifetime (27). For longer lifetimes the instrument can be
calibrated with a lifetime standard such as hexafluoro-
acetone at 100 torr and 25°C which has a lifetime of 84 ±
0.5 ns (14).

     In order to accurately measure fluorescence lifetimes
over a wide dynamic range the instrument must be properly
calibrated. The TCSP technique is normally used to measure
lifetimes in the range between 0.1 and 100 ns. Calibration
over this range can be accomplished by selecting compounds
with known lifetimes. Birks and Munro (10) list the
experimental lifetimes of several aromatic molecules.
Chen (38) has suggested that quinine-NaCl solutions be used
as standard reference lifetimes standards. The reported
fluorescence of quinine taken from the literature is
18.91 ± 0.56 ns. Quinine fluorescence is quenched by NaCl
and standard solutions of known quinine/NaCl ratios give
reproducible lifetimes in the range of 0.189 to 18.9 ns.
Similarly, γ-pyrenebutyrate/KI standard solutions have
lifetimes between 18 and 115 ns.

     The sensitivity of the TCSP technique cannot be stated
in general terms. Even very weak fluorphor decay curves
can be obtained by this method. Only 1 luminescent photon
from 99 excitation flashes is required to reach the
detector in order to build up the decay curve. Scattered
light is often many orders of magnitude more intense than
the luminescence emitting gas samples (14). The use of
excitation and emission monochromators helps eliminate the
problem, and high intensity laser exciting sources
increases the sensitivity. Knight and Selinger (4) suggest
that anthracene vapor be used as a sensitivity standard to
gauge instrumental performance.

     The time required to obtain a decay curve depends
mainly on the source repetition rate. Zimmerman and coworkers (25) report that only 5 minutes was required to
obtain decay curves having 2000 counts in the peak channel
for short-lived fluorphors. The lamp frequency was 20 kHz,

and the data collection rate was 5%. At more conservative
count rates, smaller lamp frequencies, or low quantum
yields more time is required to accumulate a well defined
decay curve.

Lewis and coworkers (18) recently published a paper
containing many helpful hints on the operation of a TCSP
instrument. The construction and operation of nanosecond
flashlamps, the installation and maintenance of PMT's, and
the general operation of the apparatus are covered.

DATA PROCESSING

A hypothetical example of a decay curve obtained by
TCSP technique is shown in Figure 6. The abcissa is the
MCA channel number which represents increments of time and
the ordinate is the single photon counts, directly proportional to the emission intensity. Flashlamp and fluorescence decay curves are shown along with the undistorted
decay law. If the fluorescence decay is a first order
rate process then a simple exponential law is followed in
the equation

$$E(t) = E(0) \exp(-kT) \tag{2}$$

where $E(t)$ and $E(0)$ are the fluorescence emission intensities at time t and at time zero, respectively, k is the
decay constant, and T is elapsed time. The reciprocal of
k is the lifetime, τ. The lifetime is defined as the time
required for the emission to decay to 1/e of the initial
intensity. The lifetime is computed by preparing a graph
of the natural logarithm of intensity versus time for the
portion of the data following the peak, obtaining the
slope of the straight line portion, and calculating the
reciprocal.

The graphical slope method is suitable for first
order decay provided the fluorescence emission curve is
undistorted by the flashlamp decay. If the fluorescence
lifetime is nearly equal to or less than the flashlamp
lifetime,then the emission curve is distorted and the
lifetime must be calculated by the mathematical process of
deconvolution. Graphical slope evaluation is sufficiently
accurate in many cases. Ware (12) shows some typical

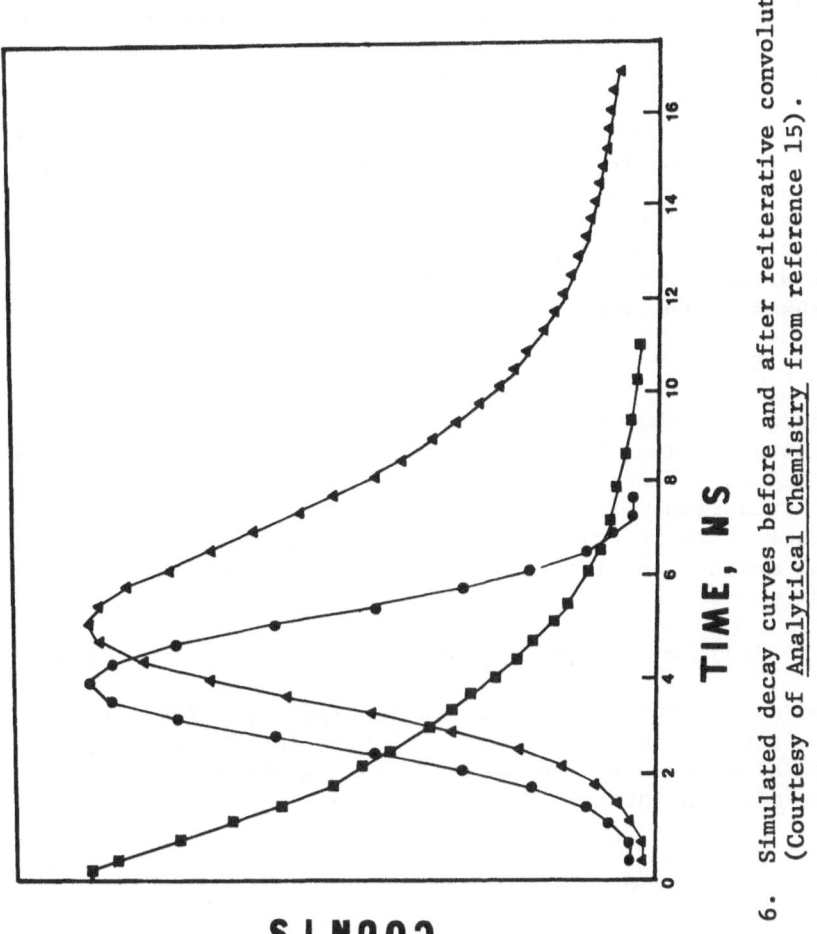

Figure 6.   Simulated decay curves before and after reiterative convolution. (Courtesy of Analytical Chemistry from reference 15).

corrections applied to calculated lifetimes where flashlamp distortion was a factor. Computer simulations in
this laboratory show as a rule-of-thumb that when graphical
slope-calculated lifetimes are within 2 ns of the flashlamp
lifetime, predictable serious errors are incurred (39).
Decay curves produced as a result of emission from mixtures
of two or more fluorophors must be deconvoluted.

In deconvolution both the flashlamp and fluorescence
decay curves are used. The flashlamp curve is considered
to be a train of infinitely narrow pulses of light called
delta function pulses. Each delta function pulse excites
singlets in the sample. The singlets then decay emitting
fluorescence photons. The first order decay law for a
single delta function pulse then becomes

$$E(t) = Ia \exp(-kT) \tag{3}$$

where I is the intensity of the pulse and a is an amplitude
constant. The flashlamp is a train of pulses such that the
emission intensity at any time E(t) equals the sum of the
contributions of singlets that were excited by all previous delta function pulses. For example the emission at
the exact time of delta function pulse three equals the
sum

$$E(3) = I_1D_2 + I_2D_1 + I_3D_0 \tag{4}$$

where D is the decay function of $\exp(-kT)$. The D sub-
scripts represent the number of time intervals over which
the singlets have decayed. This is generalized by the
convolution integral

$$E(t) = \int_0^t I_j D(t-j)dj \tag{5}$$

The emission, E(t) and the flashlamp intensity, I_j,
are experimental quantities. The decay function may be
exponential, a sum of exponentials, or some other decay
law. The purpose of deconvolution is to solve the convolution integral for a and k in Equation 3. The manipula-
tion of between 100 and 500 data points in the solution
of the convolution integral requires a digital computer.
Several mathematical algorithms have been developed and
applied to deconvolution of experimental fluorescence

decay curves. Isenberg and Dyson (40-44) have developed
and tested the method of moments. Gafni, Modlin, and
Brand (45) recently reported a deconvolution method using
Laplace transformation. Schlesinger (46) reported deconvolution by Fourier transformation, and Valeur and Moirez
(47) have developed the method of modulating functions.
Grinvald and Steinberg (48), and Ware and coworkers (49)
use the method of least squares in deconvolution. Zimmerman and coworkers (36,50) developed the reiterative convolution method. In this laboratory a computer program
was written to deconvolute decay curves by the reiterative
convolution method. The program is written in BASIC
language and was designed for use in a time-sharing PDP/11
computer. The method has been critically evaluated on both
computer-simulated and experimental data (27,39). Reiterative convolution appears to be at least the equivalent in
accuracy, precision, resolution of multiexponentials, and
flexibility as other commonly used procedures.

Reiterative convolution is a method of fitting a curve
to the decay data. Initially intelligent guesses are made
for a and k. The computer then reiteratively calculates
better values for a and k until a satisfactory fit is
obtained. Assuming fluorescence decay follows a negative
exponential, the decay function is written as the general
expression

$$D(t) = \sum_{n=1}^{n} a(n) \exp[-k(n)T(t)] \tag{6}$$

where n is the number of exponentially decaying components
present, t is an index counter on the time axis, and T is
elapsed time. Expressing the convolution integral as a
summation gives

$$E(t) = \sum_{j=1}^{t} I(j) \sum_{n=1}^{m} \left[a(n) \exp[-k(n)T(t-j+1)] \right] \tag{7}$$

The index counter, j, has been changed to begin at one
instead of zero (as in Equation 5) to make computerization
more convenient. Differentiating Equation 7 with respect
to k(n) gives

$$\frac{\partial E(t)}{\partial k(n)} = -a(n) \sum_{j=1}^{t} I(j)T(t-j+1)\exp[-k(n)T(t-j+1)] = K(t,n)$$

$$(8)$$

and with respect to a(n) gives

$$\frac{\partial E(t)}{\partial a(n)} = \sum_{j=1}^{t} I(j) \exp[-k(n)T(t-j+1)] = A(t,n) \qquad (9)$$

Note that one set of A and K values are required for each component.

Using the relation

$$dE(t) = \frac{\partial E(t)}{\partial k(n)} dk(n) + \frac{\partial E(t)}{\partial a(n)} da(n) \qquad (10)$$

the following relationship is obtained.

$$\Delta E(t) = K(t,n) \ \Delta k(n) + A(t,n) \ \Delta a(n) \qquad (11)$$

The first step in reiterative convolution is to supply the computer with the experimental values for E, I, and T. Second, intelligent guesses are made for a(n) and k(n) in Equation 7. The computer, using Equation 7, calculates a set of values $E(t)_{calc}$. The difference $[E(t)_{calc} - E(t)_{expt}]$ equals $\Delta E(t)$. When $\Delta E(t)$ is minimized over the whole decay curve then a(n) and k(n) are good approximations of the decay law values. In order to minimize $\Delta E(t)$, Equation 11 is solved for $\Delta k(n)$ and $\Delta a(n)$.

This is accomplished by matrix algebra for a single exponential in the form

$$\begin{bmatrix} \Delta E(1) \\ \Delta E(2) \\ \Delta E(3) \\ \Delta E(N) \end{bmatrix} = \begin{bmatrix} A(1,1) & K(1,1) \\ A(2,1) & K(2,1) \\ A(3,1) & K(3,1) \\ A(N,1) & K(N,1) \end{bmatrix} \begin{bmatrix} \Delta a(1) \\ \Delta k(1) \end{bmatrix} = \overline{\Delta E} = \overline{\overline{AK}}\overline{\Delta}$$

$$(12)$$

where N is the number of data points. The values for K(t,n) and A(t,n) are obtained by computer calculation of Equations 8 and 9. To solve for the $\overline{\Delta}$ vector both sides of the equation are multiplied by the transpose of the $\overline{\overline{AK}}$ matrix. Then both sides are multiplied by the inverse of the $(\overline{\overline{AK}}^T\overline{\overline{AK}})$ matrix

$$(\overline{\overline{AK}}^T\overline{\overline{AK}})^{-1}\overline{\overline{AK}}^T\overline{\Delta E} = \overline{\Delta} \tag{13}$$

The Δa and Δk values thus obtained are subtracted from the original values of a and k, respectively, and better estimates of a and k are attained. The process is repeated until the variance of the fit is not significantly different from the variance of the data.

Knight and Selinger (26) report a method for estimating the goodness of fit by comparing the statistical variance of the data to the mean variance of the fit. Ideally the ratio of the two should be unity. Deconvolution of experimental data show the variance of the fit to be about 3 times the variance of the data (36). Zimmerman and coworkers (25) suggest a more practical criterion for fit, the A value. The A value is defined as the sum of the absolute value of the point by point difference between experimental points divided by the sum of data points.

$$A = \sum_{t=1}^{N} |\Delta E(t)| \bigg/ \sum_{t=1}^{N} E(t) \tag{14}$$

An A value of less than 0.03 is considered an acceptable fit.

APPLICATIONS

The principal application of TCSP technique is the measurement of decay kinetics in the nanosecond range. Some of the same electronic modules are also used in general timing experiments, especially for nuclear radiation phenomena. Other quantum detectors such as Ge(Li) and NaI(Tl) have been used in modifications of TCSP to measure these phenomena (13). It is beyond the scope of this report to include those studies. Some applications of TCSP technique to nanosecond fluorometry will be presented and the significance of lifetime information derived from the data will be discussed.

In fluorescence studies the four main quantities that can be measured are: 1) excitation and emission spectra,

2) polarization, 3) quantum yield, and 4) fluorescence decay kinetics, which include the lifetime, τ. The measurement of τ is the most difficult because the fluorescence
decay rates are in the 10^{-12} to 10^{-6} s range and exceed the
capability of conventional instrumentation. The technological improvements in TCSP technique since its development
in 1961 has extended fluorescence decay measurements into
the nano- and subnanosecond lifetime range (19).

Molecules and atoms that are excited by absorption of
photons enter an activated state M*. Deactivation proceeds
by fluorescence, phosphorescence, or by some radiationless
process. Radiationless transitions compete with fluorescence in the depopulation of the excited state. This compe-
tition is reflected in the value of τ. The greater the
radiationless deactivation, the shorter the decay lifetime.
Fluorescence lifetime measurement is an indirect method of
obtaining information on the radiationless processes of rotation and diffusion, photochemical reactions, quenching
mechanisms, solvent interactions, energy transfer, and excited state complex formation. Chen (51) has outlined some
of the applications of lifetime measurements in biological
chemistry. Many references to specific applications of
nanosecond fluorometry are found in a review by Weissler
(52). Birks and Munro (10) have tabulated the results of
lifetime measurements of several aromatic molecules. A general review of the luminescence literature was issued by
Cundall and Palmer (53) containing many references to lifetime measurement studies. Weber (54) reviewed some recent
developments in the use of nanosecond fluorometry in the
area of biophysics. Other recent reviews of applications
in biophysics include the study of macromolecules by
Yguerabide (16), a general review of molecular interactions
and structure as analyzed by fluorometry by Rigler and Ehrenberg (55), and another on the application of fluorescent
probes in membranes authored by Azzi (56).

Fluorescence Polarization Studies

Fluorescence lifetimes are important in the study of
the structures and conformations of macromolecules
(51,54-58). Fluorescent molecules can be attached to
selected sites of macromolecules such as proteins and
enzymes. These fluorescent probes are excited by polarized light and the resulting luminescence will be polarized to some degree. Rotation of molecules due to

Brownian motion depolarizes the emission. By measuring the amount of fluorescence polarization and the fluorescence lifetime, the amount of Brownian rotation can be calculated.

The relationship between the fluorescence polarization, P, and τ, is given by the Perrin equation

$$1/P - 1/3 = (1/Po - 1/3)(1+3\tau/\rho) \tag{15}$$

where Po is the polarization observed in rigid media, and ρ is the rotational relaxation time. The ρ value is a measure of the Brownian rotational mobility and is obtained from the equation

$$1/\rho = 1/3 (1/\rho_1 + 1/\rho_2 + 1/\rho_3) \tag{16}$$

where ρ_1, ρ_2, and ρ_3 are the rotational relaxation times about the three axes of the molecular ellipsoid. The details of these measurements and calculations are found in the literature (16,51,54-57). Fluorescence depolarization studies give information on the shape of proteins. Protein relaxation time, ρ, is also a measure of the elongation of the macromolecule. The greater ρ, compared to the relaxation time of a sphere of the same molecular weight, the greater the elongation (51,57). Accurate measurement of lifetimes is important because the lifetimes of fluorescent probes vary depending on the substrate and degree of fluorescent labeling. Other studies using fluorescence depolarization have been recently reported by Hammes and coworkers (59,60) in the study of rotations of chloroplast coupling factor 1.

Fluorescence polarization measurements are also useful in detecting conformational transitions of allosteric enzymes (61,62). The commonly used fluorescent probe 1-anilino-naphthalene-8-sulfonate (ANS) was bound to phosphorylase b, a dimer consisting of two subunits each having molecular weights of approximately 92,500. The decay curve of this complex consisted of two exponentials which were deconvoluted to give lifetimes of 8.7 and 19.2 ns, respectively. This illustrates the sensitivity of probes to environment. Addition of the activators AMP or IMP, or conversion to phosphorylase a resulted in a decrease in the longer τ which indicated dissociation of ANS from the corresponding site because of a change in conformation (62). The measured polarization of the AMP-induced conformational change indicated no major change in the molecular shape and rigidity of the enzyme.

Fluorescence anisotropy decay measurements have been made by Wahl and coworkers (63,64) on ethidium bromide-DNA complexes and on ethidium bromide-poly d(A-T) complexes. Both DNA and poly d(A-T) have helix structures and fluorescence polarization studies show the ethidium bromide dye unwinds this helix structure. Conti (65) recently reviewed applications of nanosecond fluorometry to fluorescent probes attached to nerve membranes.

The media in which the molecules are placed can affect the degree of rotation. Cehelnik and coworkers (66) placed all-trans-1,6,-diphenylhexa-1,3,5-triene (DPH) in various media including liquid crystals. The molecules were specifically oriented in the liquid crystal media while rotational diffusion in isotropic media was greater.

Fluorescence Quenching Studies

Fluorescence emission is quenched by collision of excited molecules with substances such as iodide ion, chloride ion, and oxygen. The rate of quenching can be calculated if the fluorescence lifetime is measured. When a quencher, Q is added to the sample the quantum yield is obtained from the equation

$$\Phi' = \frac{k_o}{k_o + k_1 + k_q \,[Q]} \qquad (17)$$

where k_o and k_1 are the rates of the emission and radiationless process, k_q is the quenching rate, and $[Q]$ is the concentration of quencher. In absence of quencher the quantum yield is given by

$$\Phi = \frac{k_o}{k_o + k_1} \qquad (18)$$

the ratio of Equations 18 and 19 is given below.

$$\Phi/\Phi' = 1 + \frac{k_q}{k_o + k_i} [Q] \qquad (19)$$

The Stern-Volmer constant, K_{sv}, is given by

$$K_{sv} = \frac{k_q}{k_o + k_i} = k_q \tau \qquad (20)$$

Then substituting Eq. 20 into Eq. 19 one obtains

$$\Phi/\Phi' = 1 + K_{sv} [Q] \qquad (21)$$

K_{sv} is obtained from a plot of Φ/Φ' versus [Q]. If the decay lifetime, τ, in absence of quencher is known then k_q can be calculated. Values of k_q can be used to deduce collision rates (51,67). Chen (38) used Stern-Volmer plots to obtain K_{sv} and k_q values for solutions of quinine quenched by NaCl and solutions of γ-pyrenebutyrate quenched by KI. The fluorescence lifetimes of standard solutions of known fluorophor/quencher ratios were then calculated and used as lifetime standards to calibrate TCSP instrumentation.

Photochemical Studies

Fluorescence lifetime data are useful in helping deduce the mechanisms of organic photochemical reactions. Zimmerman and coworkers (36,68) utilized TCSP to measure the rates of rearrangement of irradiated acyclic di-π-methane compounds. Room temperature rearrangement rates are exceedingly rapid and are of the order of 10^{10} to 10^{12} s^{-1}. These photochemical reactions are luminescent processes and the rates of disappearance of excited singlets is a direct measure of the rates of rearrangement. The approach used to obtain the picosecond decay rates was to first measure the slower decay lifetime at 77°K. Then the fluorescence quantum yields at 77° K and at room temperature were measured. A "magic multiplier", M, is calculated from the ratio of the two quantum yields.

$$M = \Phi_{77}/\Phi_{rt} \qquad (22)$$

M is used to calculate the excited state decay at room temperature (36). This method. extends the rate measurement capability to 10^{12} s^{-1}.

In a more recent study (25) fluorescence lifetimes were used to study the excited state twisting of a series of 1-phenylcycloalkenes having between 4 and 8-membered rings. The twisting of these cyclic compounds can create variations in the singlet energy gaps and subsequent variations in excited state decay lifetime.

Energy Transfer Studies

Under some conditions energy can be transferred from excited molecule M* to an unexcited molecule M' in the fashion

$$M* + M' \longrightarrow M + M'* \tag{23}$$

Singlet-singlet energy transfer is important in biochemical systems. This has been called radiationless transfer, dipole-dipole resonance transfer, Förster-type energy transfer, long-range non-radiative transfer and, if M'* is luminescent, sensitized fluorescence. The critical distance at which transfer occurs is partly a function of the fluorescence lifetime (51).

The energy transfer can be observed in four ways: 1) the donor molecule is fluorescent and energy transfer results in quenching, 2) the acceptor molecule is fluorescent and energy transfer results in sensitized fluorescence, 3) like molecules tend to transfer energy and depolarization of fluorescence results, 4) a quenching agent acts to dissipate the energy transfer and greater than normal quenching results.

Some recent applications of nanosecond fluorometry in energy transfer studies include work by Loper and Lee (69) on the steric hindrance of gas-phase singlet energy transfer from naphthalene to the trans-azobutane isomers. Other studies by Seibert, Alfano and Shapiro (70) report the possible energy transfer between carotenoids or chlorophyll b to chlorophyll a.

Other Applications

Chen (51) notes that fluorescence lifetimes can be determined even when the sample is turbid. This type of sample produces severe scattered light which changes the emission intensity and quantum yields. Fluorescence lifetimes are proportional to quantum yields and are not influenced by scatter.

Also, fluorescence lifetimes aid the elucidation of quenching mechanisms. If quencher-fluorophor complexes are formed in the ground state, then the lifetime is independent of quencher concentration. However, if the complex is formed in the excited state then τ is inversely proportional to quencher concentration because the rate of fluorescence is increased by the increased quenching rate (51). Nemzek and Ware (71) studied the kinetics of diffusion-controlled processes and specifically transient effects in fluorescence quenching. They predict that quenched fluorescence to follow the decay law

$$D(t) = \exp(-at - 2b\sqrt{t})$$ (24)

where a and b are constants and $D(t)$ is the decay function at time t.

Locken and coworkers (72) used the TCSP technique to measure the rates of excited state proton transfer reactions in compounds such as 2-napthol. Ionization constants of organic acids in the excited state may differ by several orders of magnitude from those of the ground state.

Grinvald and Steinberg (73) studied the decay kinetics of tryptophan and discussed the implications of observed nonexponential decay in this system. Churchich and coworkers (74) utilized nanosecond fluorometry to study nonequivalent binding sites of cystathionase.

Hanson and Lee (75,76) and Loper and Lee (77) used the TCSP technique to determine lifetimes of molecules in the gas phase and compared the experimental lifetimes with those calculated from theoretical considerations.

SUMMARY AND CONCLUSIONS

Fluorescence phenomena occuring in the nanosecond range are measured in order to glean information about the structure and function of molecules. The time correlated single photon technique has proven a valuable tool for obtaining the temporal characteristics of molecular species. Lifetime information is particularly helpful in the study of biological macromolecules such as membranes, proteins, and enzymes. These are very complex species and the processes in which they participate occur in the nano- and picosecond ranges. Many of the applications listed above also require resolution in these time ranges and the use of the TCSP technique is increasing as investigators seek to probe these fast phenomena.

A current research report (78) indicates that the use of mode-locked lasers capable of generating intense pulses with picosecond duration will increase. The use of laser sources permits measurements of lifetimes in the picosecond range and should encourage the development of new applications.

References

1. M.L. Franklin, G. Horlick,and H.V. Malmstadt, Anal. Chem., 41, 2 (1969).

2. H.V. Malmstadt, M.L. Franklin, and G. Horlick, Anal. Chem., 44, No. 8, 63A (1972).

3. G.A. Morton, App. Opt., 7, 1 (1968).

4. ORTEC, Inc., Bulletin on Model 5C1 Photon Counting System, Oak Ridge, TN. (1973).

5. M.R. Zatzick, SSR Instruments Co., Applications Note 71021, Santa Monica, Ca.

6. S.A. Miller, Rev. Sci. Instrum., 39, 1923 (1968).

7. D.O. Cooke, R.M. Dagnall, B.L. Sharp, and T.S. West, Spectros. Lett., 4, 91 (1971).

. M.K. Murphy, S.A. Clyborn, and C. Veillon, <u>Anal. Chem.</u>, <u>45</u>, 1468 (1973).

. M.R. Zatzick, Research/Development, Nov. 1970.

0. J.B. Birks and I.H. Munro, "Progress in Reaction Kinetics", G. Porter, ed., Pergamon Press, New York, 1967, Vol. 4, Chapter 7.

1. W.R. Ware, Office of Naval Research, Contract N00014-67-A-0113-0006, Technical Report No. 3 (1969).

2. W.R. Ware, "Creation and Detection of the Excited State", A.A. Lamola, ed., Marcel Dekker, Inc., New York, 1971, Vol. 1, Part A, Chapter 5.

3. ORTEC, Inc., Application Note AN 35, Oak Ridge, TN. (1971); ORTEC, Inc., Life Science Notes, No. 1, Oak Ridge, TN. (1974); ORTEC, Inc., 9200 nanosecond Fluorescence Spectrometer Descriptive Literature, Oak Ridge, TN. (1972).

4. A.E.W. Knight and B.K. Selinger, <u>Aust. J. Chem.</u>, <u>26</u>, 1 (1973).

5. L.J. Cline Love and L.A. Shaver, <u>Anal. Chem.</u>, 48, No.4, A pages (1976).

6. J. Yguerabide, "Methods in Enzymology", S.P. Colowick and N.D. Kaplan, eds., Academic Press, New York, 1972, Vol. 26, Part C.

7. J.N. Demas and G.A. Crosby, <u>Anal. Chem.</u>, <u>42</u>, 1010 (1970).

8. L. Hundley, T. Coburn, E. Garwin, and L. Stryer, <u>Rev. Sci. Instrum.</u>, <u>38</u>, 488 (1967).

9. L.M. Bollinger and G.E. Thomas, <u>Rev. Sci. Instrum.</u>, 32, 1044 (1961).

0. Y. Koechlin, Ph.D. Thesis, University of Paris, 1961.

1. Y. Koechlin, <u>C.R. Acad. Sci.</u>, (Paris), <u>252</u>, 391 (1961).

22. P.B. Coates, J. Phys. E., 5, 148 (1971).

23. C.C. Davis and T.A. King, Rev. Sci. Instrum., 41, 407 (1970).

24. D.E. Donohue and R.C. Stern, Rev. Sci. Instrum., 43, 791 (1972).

25. H.E. Zimmerman, K.S. Kamm, and D.P. Wertheman, J. Amer. Chem. Soc., 97, 3718 (1975).

26. A.E.W. Knight and B.K. Selinger, Spectrochim. Acta, 27A, 1223 (1971).

27. L.A. Shaver and L.J. Cline Love, Unpublished data, 1975.

28. J. Zynger and S.R. Crouch, App. Spectros., 26, 631 (1972).

29. F.E. Lytle and M.S. Kelsey, Anal. Chem., 46, 855 (1974).

30. E.F. Wyner, J.A. Sousa, and J.F. Roach, Spectros. Lett. 8, 419 (1975).

31. P.M. Rentzepis and C.J. Mitschele, Anal. Chem., 42, (No. 14), 21A (1970).

32. V.H. Kollman, S.L. Shapiro, and A.J. Campillo, Biochem. Biophys. Res. Commun., 63, 917 (1975).

33. D.J. Bradley, Contemp. Phys., 16, 263 (1975).

34. V.V. Arsen'ev, V.A. Gavanin, V.Z. Pashchenko, S.P. Protasov, L.B. Rubin, and A.B. Rubin, J. Appl. Spectros., 18, 801 (1973).

35. W.M. Watson, Y. Wang, J.T. Yardley, and G.D. Stucky, Inorg. Chem., 14, 2374 (1975).

36. H.E. Zimmerman, D.P. Werthemann, and K.S. Kamm, J. Amer. Chem. Soc., 96, 439 (1974).

37. C. Lewis, W.R. Ware, L.J. Doemeny, and T.L. Nemzek, Rev. Sci. Instrum., 44, 107 (1973).

38. R.F. Chen, Anal. Biochem., 57, 593 (1974).

39. L.A. Shaver and L.J. Cline Love, Appl. Spectros., 29, 485 (1975).

40. I. Isenberg and R.D. Dyson, Biophys. J., 9, 1337 (1969).

41. R.D. Dyson and I. Isenberg, Biochem.,10, 3233 (1971).

42. I. Isenberg, R.D. Dyson, and R. Hanson, Biophys. J., 13, 1090 (1973).

43. I. Isenberg, J. Chem. Phys., 59, 5696 (1973).

44. I. Isenberg, J. Chem. Phys., 59, 5708 (1973).

45. A. Gafni, R.L. Modlin, and L. Brand, Biophys. J., 15, 263 (1975).

46. J. Schlesinger, Nucl. Instrum. Methods, 106, 503 (1973).

47. B. Valeuer and J. Moirez, J. Chim. Phys., 70, 500 (1973).

48. A. Grinvald and I.Z. Steinberg, Anal. Biochem., 59, 583 (1974).

49. W.R. Ware, L.J. Doemeny, and T.L. Nemzek, J. Phys. Chem., 77, 2038 (1973).

50. H.E. Zimmerman and T.P. Cutler, Chem. Commun., 1975, 598 (1975).

51. R.F. Chen, Fluorescence News, 8, 29 (1974), American Instrument Co., Silver Spring, Md.

52. W. Weissler, Anal. Chem., 46, 500R (1974).

53. R.B. Cundall and T.F. Palmer, "Annual Reports on the Progress of Chemistry", The Chemical Society, London, 1973, Vol. 70, Section A. Chapter 3.

54. G. Weber, Ann. Rev. Biophys. Bioeng., 1, 553 (1972).

55.   R. Rigler and M. Ehrenberg, Quart. Rev. Biophys., 6,
      139 (1973).

56.   A. Azzi, Quart. Rev. Biophys., 8, 237 (1975).

57.   T. Tao, Biopolymers, 8, 609 (1969).

58.   L. Stryer, Science, 162, 526 (1968).

59.   L.C. Cantley, Jr. and G.G. Hammes, Biochem., 14,
      2976 (1975).

60.   S. Matsumoto and G.G. Hammes, Biochem., 14, 214 (1975).

61.   M.S. Tung and R.F. Steiner, Biochem. Biophys. Res.
      Commun., 57, 876 (1974).

62.   M.S. Tung and R.F. Steiner, Bipolymers, 14, 1933 (1975).

63.   D. Genest, Ph. Wahl, and J.C. Auchat, Biophys. Chem.,
      1, 266 (1974).

64.   J.L. Tichadon, D. Genest, Ph. Wahl, and G. Aubel-
      Sadron, Biophys. Chem., 3, 142 (1975).

65.   F. Conti, Ann. Rev. Biophys. Bioeng., 4, 287 (1975).

66.   E.D. Cehelnik, R.B. Cundell, J.R. Lockwood, and
      T.F. Palmer, J. Chem. Soc., Faraday Trans. II, 70,
      244 (1974).

67.   S.S. Lehrer, Biochem., 10, 3254 (1971).

68.   H.E. Zimmerman, D.P. Werthemann, and K.S. Kamm,
      J. Amer. Chem. Soc., 95, 5094 (1973).

69.   G.L. Loper and E.K.C. Lee, J. Chem. Phys., 63, 3779
      (1975).

70.   M. Seibert, R.R. Alfano, and S.L. Shapiro, Biochem.
      Biophys. Acta, 292, 493 (1973).

71.   T.L. Nemzek and W.R. Ware, J. Chem. Phys., 62, 477
      (1975).

72. M.R. Locken, J.W. Hayes, J.R. Gohlke, and L. Brand, Biochem., 11, 4779 (1972).

73. A. Grinvald and I.Z. Steinberg, Biochem., 13, 5170

74. J.E. Churchich, T. Beeler, and K. Ja Oh, J. Biological Chem., 250, 7722 (1975).

75. D.A. Hanson and E.K.C. Lee, J. Chem. Phys., 62, 183 (1975).

76. D.A. Hanson and E.K.C. Lee, J. Chem. Phys., 63, 3272 (1975).

77. G.L. Loper and E.K.C. Lee, Chem. Phys. Lett., 13, 140 (1972).

78. J.L. Marx, Science, 188, 1002 (1975).

LUMINESCENCE TECHNIQUES IN DRUG ANALYSIS

J. Arthur F. de Silva
Dept. of Biochemistry and Drug Metabolism
Hoffmann-La Roche Inc.
Nutley, New Jersey 07110

I. Introduction

Studies on the pharmacokinetics of a drug, i.e., absorption, distribution, metabolism, and elimination, and the correlation of therapeutic effectiveness with blood concentration requires a suitably sensitive and specific means of chemical analysis. The development of potent drugs has significantly reduced the therapeutic doses required for the maintenance of clinical efficacy. The quantitation of blood concentrations resulting from these low doses demands highly sophisticated analytical methods capable of accurate and precise quantitation with nanogram (10^{-9} g) to picogram (10^{-12} g) sensitivity (1). The majority of the drugs in clinical use today are organic molecules containing aromatic "benzenoid" or heterocyclic ring systems containing specific functional groups which impart intrinsic analytical properties to the compound.

Due to their "aromaticity" in general, drug molecules absorb ultraviolet energy readily and consequently undergo deactivation by fluorescence or phosphorescence emission in their return to the ground state (2, 3).

From an analytical point of view, both fluorescence and phosphorescence techniques are

very sensitive and capable of quantitation in the
sub parts per million (ppm) to parts per billion
(ppb) range. They are quite specific for the
compounds being analyzed, since two kinds of
spectra (excitation and emission) are obtained
for the characterization of different compounds.

The separation in excitation and/or emission
wavelengths has enabled the simultaneous determination of multicomponent mixtures. Phosphorescence has an added dimension in selectivity,
in that the lifetimes (τ) of different species
in a mixture are different and can impart further
specificity to the technique.

A. Luminescence spectrophotometry deals with
the measurement of energy emitted by a molecule
in the excited state returning to the ground
state through a series of electron transitions
(2,3,4). The type of electron transition involved
and the lifetime of the excited species determines
the nature of the luminescence process. There are
two main types of luminescence processes: (A)
Fluorescence which involves a singlet to singlet
transition (electron spins paired) in which the
lifetime of the excited state is of the order of
10^{-8} secs, and (B) Phosphorescence which involves
a triplet to singlet transition in which the lifetime of the excited state is of much longer duration; ranging from milliseconds to seconds.

In luminescence analysis one can utilize
the intrinsic fluorescence or phosphorescence
properties of the parent molecule dissolved in a
suitable solvent or one can utilize the luminescence properties of a suitable chemical derivative of the parent molecule to produce a "fluorophor" or a "phosphor" (5). Luminescent derivatives may be produced through a number of chemical
reactions, e.g., hydrolysis, condensation, photochemical reactions, ring closure, or cyclization,
coupling to another compound, molecular rearrangement, dehydrogenation, oxidation, and complex

formation. All these techniques have been useful
in preparing derivatives for fluorescence and
phosphorescence analysis of drugs (6).

B. Molecular structure and luminescence. In
general, most aliphatic compounds neither absorb
in the near UV or visible regions nor luminesce
except for a few long chain conjugated molecules
(carotenoids). Most aromatic compounds that
luminesce do so because they have strong UV
absorption properties which tend to produce
excited states readily. Consequently, inspection
of the chemical structure and the functional
groups on the molecule would enable one to predict with a reasonable degree of certainty the
probablility that a compound in solution might
fluoresce. Aromatic ring systems (benzenoid or
heterocyclic indoles, isoquinolines) are "π"
electron rich systems and have a high resonance
energy. Depending on the type of functional
groups and substituents on the aromatic ring
system, they tend to absorb radiation giving rise
to well defined UV or visible range spectra (7).

The following general properties that
influence the electron density and resonance
energy of the molecule are conducive to increasing the UV absorption and the luminescence
properties of aromatic compounds: (a) fusion of
the benzene ring to other benzene rings, e.g.,
naphthalene, anthracene, and phenanthrene, or to
heterocyclic rings as in quinoline, isoquinoline,
and indoles to produce a rigid planar molecule;
(b) certain electron-donating functional groups
on the aromatic nucleus, such as amino, dimethyl-
amino, phenol, alkyl chains, hydroxy, methoxy
carbonyl, and nitrile groups, all tend to
increase UV absorption and the luminescence of
the molecule; and (c) certain electron-withdraw-
ing functional groups, such as the halogens,
carboxyl, and nitro groups tend to decrease UV
absorption and the luminescence of the molecule
(8, 9).

Most compounds of biological and pharmaco-
logic interest are ionizable and can exist in
aqueous solution at various stages of dissociation depending on the prevailing pH of the medium.
The relationship between molecular dissociation,
the nature of the luminescent species (fluorophors), and the pH of the medium is shown in
Figure 1, for 2-hydroxynicotinic acid, a hypo-
cholesteremic agent (10).

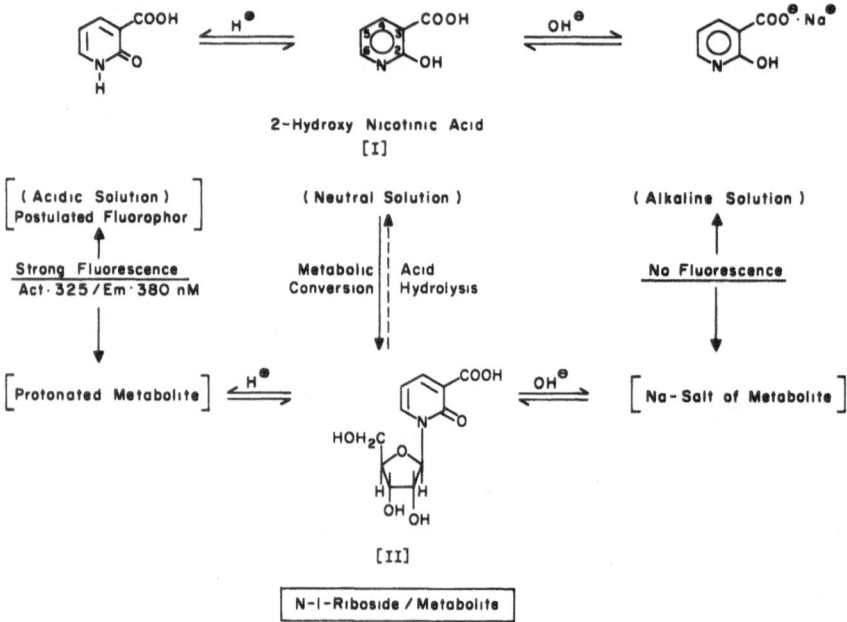

Figure 1.   The influence of pH on the luminescence
of a compound.

In neutral solvents, the parent compound
shows weak fluorescence which is markedly
enhanced in acidic media probably due to the
formation of the tautomeric ketone. The major
metabolite, the N-1-riboside, however shows
strong intrinsic fluorescence under neutral and
acidic pH since the compound exists as the ketone.
Neither compound fluoresces in alkaline media due
to the formation of the Na+ salt.

C. Screening for intrinsic luminescence behavior

A preliminary examination of the luminescence properties of compounds can be made on thin layer chromatoplates at room temperature (298K) and at liquid nitrogen temperature (77K). Samples of 100 µg, 1 µg, and 0.1 µg are applied to each of three channels scored on a 20 x 10 cm thin layer chromatoplate which are then sprayed lightly with methanolic 0.1 N NaOH, methanol, and methanolic 0.1 N HCl respectively to cover the essential pH range. The plate is then photographed with a polaroid camera using a Tiffen (Hi-Trans, Yellow 1, Series 7) filter, first in visible light, then in shortwave ultraviolet light (254 nm), and at room temperature. Any intrinsic UV or visible range absorption or fluorescence behavior observed is recorded.

The plate is then resprayed with the respective mixtures until completely saturated, placed in an aluminum pan, and sufficient liquid nitrogen is poured into the container, to completely submerge the plate. The plate is observed again under shortwave (254 nm) irradiation, and any cryogenic fluorescence observed is photographed. The excitation energy is then switched off, and any phosphorescence emission is photographed immediately thereafter. The color and duration of the visible emission are also recorded. In order to photograph the phosphorescence, multiple exposures at wide aperature settings are necessary. Observations of fluorescence and phosphorescence emission are made again under long-wave (366 nm) irradiation. This simple procedure provides valuable insight into the intrinsic luminescence behavior of a compound.

The use of thin layer chromatography as a rapid qualitative procedure for the evaluation of luminescence properties at low temperatures (77K) (11,12,13) significantly facilitates the selection of the analytical parameters for low temperature luminescence studies. The sensitivity and selectivity of detection of complex mixtures of organic compounds is also enhanced.

Fluorescence and phosphorescence spectra of compounds showing luminescence are readily obtained with a suitable spectrofluoro or phosphorimeter in the solvent mixture whose pH is optimal for generating luminescence emission. Linear concentration curves are also determined to cover the useful range of quantitation.

II. Practical aspects of luminescence techniques

A. Instrumentation. Fluorescence measurements are made in filter photofluorometers or spectro-photofluorometers designed to cover the wavelength region from 200 nm to 800 nm in both excitation and emission monochromators. Typical spectrophotofluorometers commercially available (Figure 2) are equipped with double grating mono-chromators, a Xenon arc energy source, a photo-multiplier tube to record the signal, and use a right angle geometry in the sample cell for excitation of the sample with the luminescence emitted measured at right angles (14).

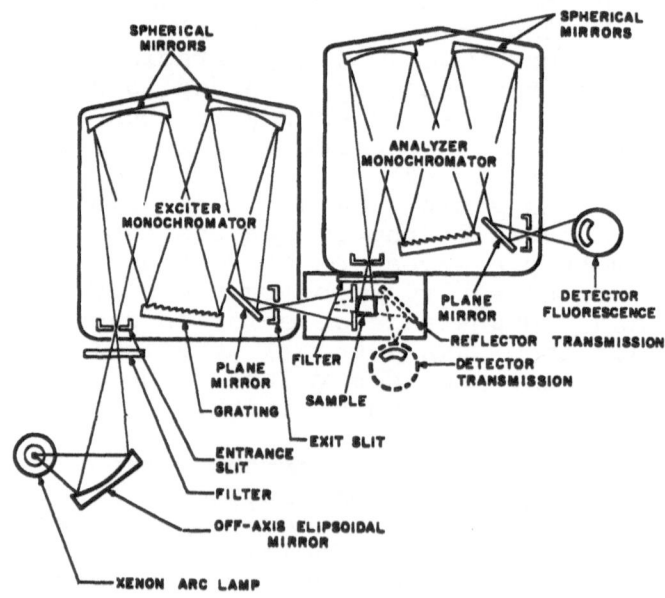

Figure 2.   The optical path of the Farrand Mark I Spectrofluorometer.

Phosphorescence measurements are made in a
spectrofluorometer modified to accommodate a
phosphoroscope as illustrated in Figure 3. The
major modifications consist of a mechanical
chopper device which permits intermittent excitation of the sample similar to a pulsing technique
with a built-in time factor. This enables all
fluorescence to decay and the phosphorescence
emission produced to be measured during the time
period when the sample is not being excited. A
second modification is in the sample chamber,
which consists of a micro Dewar flask to hold the
liquid nitrogen and which has a transparent
window in the silvered walls to permit the excitation energy to reach the sample and the

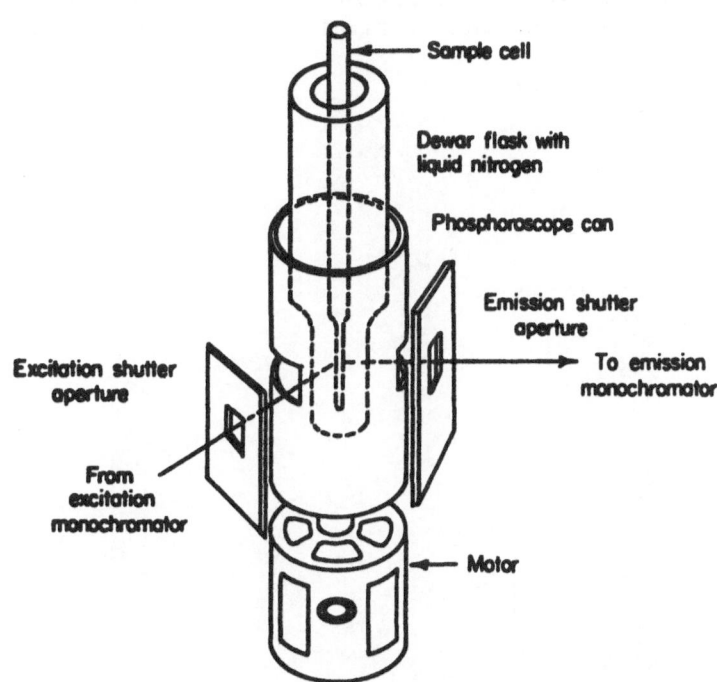

Figure 3. Components of the Lewis-Kasha phosphoroscope.

phosphorescence emission to be directed into the
emission monochromator. Special quartz sample
tubes are required to freeze the sample at 77°K
to a rigid glass for phosphorescence measurements.
These modifications can be made with commercially
available attachments (15).

A recent report describes the modification
of a Farrand Spectrofluorometer to accommodate
a commercially available phosphoroscope (16).
The instrument was equipped with a rotary chopper
at the excitation source and an optical beam condensing system focused onto the micro cell in the
sampling area which was modified to accommodate
a micro-sample phosphoroscope without any altera-
tion of the main instrument.

The phosphoroscope and its installation in
the Farrand spectrofluorometer is shown in Figure
4 and is now available from the manufacturer.

Figure 4. Phosphoroscope assembly for the
Farrand Mark I Spectrofluorometer.

B. Quantitation. All photoluminescence processes persist for some finite period of time after cessation of the excitation energy (2, 3). Both fluorescence and phosphorescence intensities of excited molecules decay in accord with first order kinetics. The process, analogous to radioactive decay, is expressed by equation 1.

Lifetime of the excited species

$$I = I_0 \, e^{-t/\tau} \qquad\qquad \text{Equation 1}$$

where: I = fluorescence or phosphorescence intensity at time t, I_0 = maximum luminescence intensity at t = 0, t = time elapsed after removing the excitation source, and τ = mean lifetime of the excited state. Thus, experimentally, τ can be measured by recording the corrected fluorescence or phosphorescence intensity as a function of time. The lifetime of a luminescent molecule is given by the time required for the intensity to fall to I/e of its initial value. Hence fluorescence and phosphorescence decay curves are exponential in character.

Quantum efficiency of luminescence processes

The quantum efficiency or yield, $\emptyset$ is the ratio of the total energy emitted by a luminescent molecule per quantum of energy absorbed. The relationship is expressed as

$$\emptyset = \frac{\text{number of quanta emitted}}{\text{number of quanta absorbed}} = \text{quantum yield}$$

$$\text{Equation 2}$$

The higher the value of $\emptyset$, the stronger the luminescence of the compound. The quantum efficiency of a luminescent molecule is dependent on several factors, such as the intensity of the excitation source, the wavelength of excitation, the relative concentrations of the excited singlet state or triplet state species produced relative to the number of molecules in the ground state, and the rate constants for these transitions as defined by the Law of Mass Action (2,3,4).

Relationship between luminescence intensity and concentration

The fundamental equation describing this relationship is

$$I_F \text{ or } I_p = I_0 \left[1 - e^{-\varepsilon bc}\right] \cdot \left[\emptyset\right] \qquad \text{Equation 3}$$

where

I_F or I_p = total fluorescence or phosphorescence intensity emitted

I_0 = Intensity of excitation energy source

ε = Molar absorptivity of compound

c = Molar concentration of the solution

b = Optical path length of solution in cms

$\emptyset$ = Quantum efficiency (yield) of the luminescence process

For dilute solutions the equation reduces to a form analogous to the Beer-Lambert relationship represented by:

$$I_F \text{ or } I_p = I_0 \, e^{-kbc} \cdot \emptyset \qquad \text{Equation 4}$$

C. Characteristics of luminescence methods

Luminescence processes adhere to the Lambert-Beer relationship in relatively dilute solutions only where significant amounts of the incident energy are not absorbed. The luminescence produced is proportional to concentration over a range in which less than 5% of the incident energy is absorbed, here one is measuring luminescence in the absence of significant absorption.

With the sensitive instruments commercially available, quantitative luminescence measurements can be made over a wide linear dynamic range from

10^{-12} g to 10^{-6} g or greater. Thus, analytical
curves of luminescence intensity versus concentration are more conveniently drawn as a log/log
plot in order to cover a wide linear dynamic
range. This is a useful technique, in that one
can determine the concentration at which luminescence quenching occurs and also obtain the useful limit of sensitivity (17).

In addition, the following practical considerations are significant:

(i) High sensitivity. They are several orders of
magnitude more sensitive than absorptiometric
methods because of the direct measurement of the
energy emitted which is at right angles to the
incident energy and is thus resolved from it.
They have a wider linear dynamic range, because
very dilute solutions can be measured. Depending
on the luminescence quantum yield, concentrations
ranging from picograms to micrograms/ml of
solution may be determined (14).

(ii) High specificity. The excitation and
emission spectra help to characterize a compound.
Phosphorescence decay lifetimes (τ) are also
characteristic of the excimer and impart further
specificity to the technique.

(iii) Limitations. They are sensitive to pH,
temperature, purity, and chemical nature of
solvent medium, and the chemical and photochemical stability of the fluorophor. Phosphorescence
also requires the elimination of collisional
deactivation due to the solvent by the use of a
rigid transparent "glass" produced by freezing
the solvent (absolute ethanol or EPA at 77°K).
This introduces complications in measurement
which may detract from its usage in routine
analysis. These limitations can, however, be
readily circumvented (15).

D. Corrected spectra

The measurement of the "absolute fluorescence

intensity" or "fluorescence quantum yield" of a
compound (analogous to molar absorptivity in
absorption analysis) requires expensive and carefully calibrated instrumentation capable of
recording energy corrected spectra in which
spectral distortions due to instrumental arti-
facts are eliminated (18,19,20). These instruments use either a heat source (thermocouple) or
a luminescent compound of high quantum yield
(rhodamine B) as a reference in correcting for
source energy changes. One advantage of such
instruments is their ability to record the true
absorption (excitation) spectrum of compounds
present at very low concentrations, such as trace
impurities or drug metabolites.

        Energy corrected instruments are not usually
employed for routine quantitative analysis. Consequently, analytically valid fluorescence measurements have to be made with reference to some ar-
bitrarily chosen reference standard, such as
riboflavin or quinine sulfate. The latter compound is at present one of the most widely used
reference standards because of its chemical
stability in aqueous 0.1 N H_2SO_4 and its relatively high fluorescence quantum yield [0.55] (21).
Calibration of the excitation energy from the
Xenon lamp source versus the fluorescence yield
recorded by the photomultiplier/microammeter
combination of a spectrofluorometer (using a
reference standard, such as quinine sulfate) is
required for each day's analysis to maintain the
instrument at peak sensitivity and to obtain
reproducible analytical data.

III. Analysis of drugs in biological fluids

A. Extraction of drugs

        The majority of drugs in clinical use today
are aromatic molecules whose functional groups
impart specific physicochemical properties to the
compound. Based on their pKa values and solubility in organic solvents, these molecules lend
themselves to selective extraction and separation

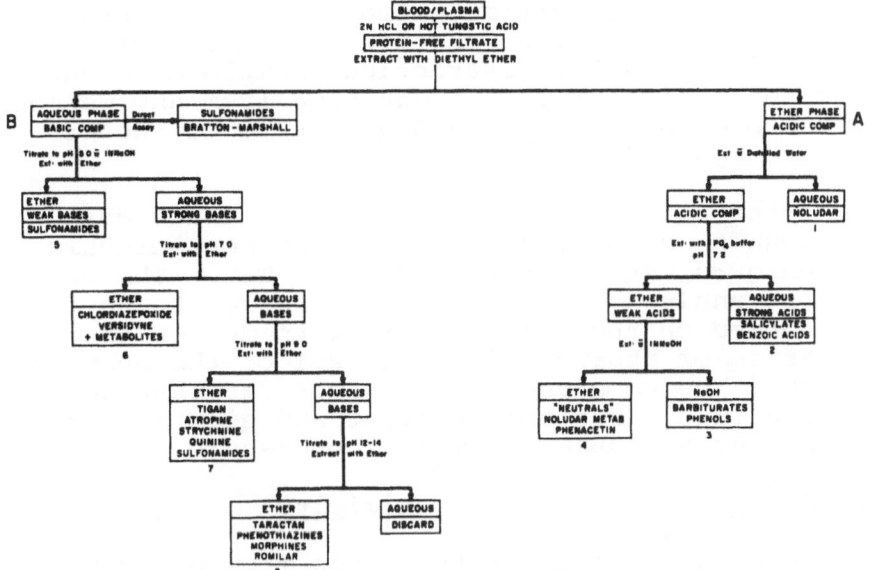

Figure 5. Systematic analysis of drugs in
biological fluids.

into three broad classes of compounds, i.e.,
acidic, neutral, and basic as shown in Figure 5.

Such a procedure is routinely used in
forensic toxicological analysis where multiple
drug ingestion is commonly involved. A protein
free filtrate (PFF) of blood or plasma is extracted with diethyl ether or chloroform to
remove the acidic and neutral drugs, leaving the
basic compounds in the aqueous medium. Sequential
extraction of the organic phase with buffers of
increasing basicity (pH) and molarity separates
the strong acids from the weak acids and neutral
molecules. Sequential titration of the aqueous
acidic phase to pH 5, 7, 9, and 13 followed by
ether extraction at each step separates the weak
bases from the intermediate and strong bases.
Quaternary compounds and those with amphoteric
behavior have to be extracted using ion-pair com-
plexation methods (22). Each of these fractions
is then analyzed further for their individual
components.

Therefore, solvent extraction at different
pH values can also be an effective means of
obtaining selective extraction, sample clean-up,
and separation of the parent drug from any metabolites present, or separate a mixture of two or
more drugs.

Further fractionation into the individual
compounds may be achieved using chromatgoraphic
techniques, such as GLC, HPLC, or TLC. The use
of TLC analysis enables a rapid screen to be
performed on the sample extract to detect the
presence or absence of any or all of the suspected
drugs. The compounds of interest can then be
eluted into a suitable solvent and quantitated
using any suitable analytical procedure.

B. Sample clean-up and specificity

The separation of the drugs from endogenous
interfering materials using a chromatographic
step such as TLC increases the specificity and
sensitvity of luminescence assays. However, in
each type of assay the limiting factor for sensitivity is the luminescence blank obtained from
eluting silica gel. Materials such as the binder
or the fluorescence indicator can be eluted,
suspensions of small amounts of colloidal silica
in the eluting solvent can increase the blank
significantly. Phosphorimetry is somewhat less
susceptible to matrix interferences than is
fluorometry, because phosphorescence is a delayed
phenomenon, whereas light scattering is instantaneous. Also, endogenous phosphorescent contaminants are less common than fluorescent
impurities. Nevertheless, TLC separation of
biological extracts is a simple, efficient, and
rapid method for effecting sample clean-up and
ensuring assay specificity, and is especially
useful in luminescence methods.

IV. Application of fluorescence analysis to drugs

Fluorometric analysis is performed in solution for the most part at ambient temperature

(room temperature 25°C). However, a significant
increase in sensitivity is attainable by cooling
the sample down to temperatures of -100 to 125°C
(temperature of acetone/dry ice or other cooling
mixtures), mainly due to a reduction in solvent
interaction with the excited species resulting in
a reduction in competitive processes which tend
to reduce the quantum efficiency of fluorescence.
A wide variety of drugs have been analyzed by
fluorescence (23) of which a few examples will
be presented.

A. Analysis using the intrinsic fluorescence of
   the parent molecule and/or its metabolites

The intrinsic or native fluorescence of an
organic compound or drug molecule has been used
to analytical advantage in their determination in
pharmaceutical formulations and in biological
fluids (24). Compounds, such as acetyl salicylate,
barbiturates, lysergic acid diethylamide (LSD),
thioxanthenes, and phenothiazines have been
analyzed using their intrinsic fluorescence behavior for quantitation.

The analysis of salicylates and barbiturates,
Figure 6, is based on their fluorescence in alkaline media.

Acetyl salicylic acid (Aspirin) has weak
intrinsic fluorescence and is rapidly hydrolyzed
chemically and biologically in acidic media to
salicylic acid; the major metabolite in blood
(Figure 6-A). The latter exhibits strong fluorescence in basic media (borate buffer pH 11.0)
with excitation at 310 nm and emission at 435 nm
which is used in its quantitation in biological
fluids (25).

The fluorescence of the barbiturates (Figure
6-B) in pH 11 to 13 buffer is due to their eno-
lization in base to form di-anionic fluorescent
species (26). The thiobarbiturates can be distinguished from the oxybarbiturates due to their
differences in their respective excitation (265

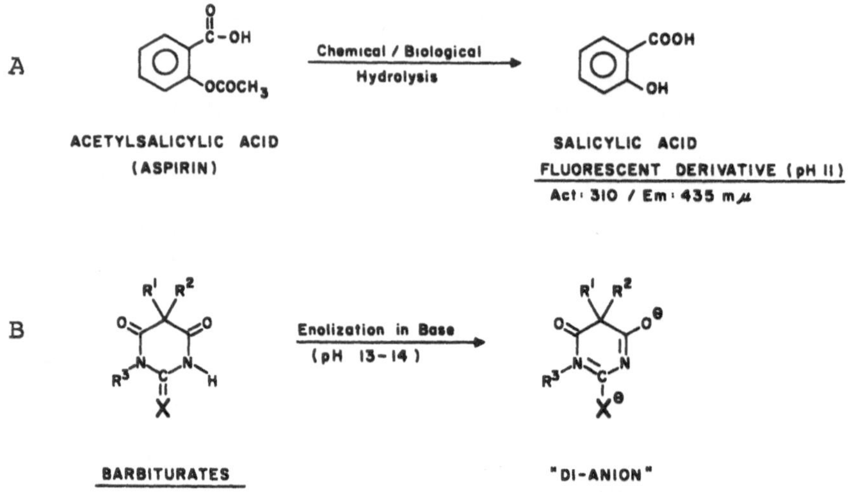

	R^1	R^2	R^3	X	Act: mμ	Em: mμ
Phenobarbital	= C_2H_5	Phenyl	H	O	265	440
Pentobarbital	= C_2H_5	1-methylbutyl	H	O	265	440
Amobarbital	= C_2H_5	Isoamyl	H	O	265	440
Thiopental	= C_2H_5	1-methylbutyl	H	S	315	530

Figure 6. Luminescence behavior of (A) Salicylates and (B) Barbiturates.

vs. 315 nm) and emission (440 vs. 530 nm) maxima.

Lysergic acid diethylamide (LSD) is a potent hallucinogen and is effective in doses as low as a few micrograms/kg of body weight. The intense intrinsic fluorescence of LSD in acidic media (0.1 N HCl) exhibiting excitation-emission maxima at 325 and 445 nm respectively was used to measure the drug in humans at a sensitivity of a few nanograms/ml of blood (27). Blood levels of the drug following an intravenous dose of 2 μg/kg of body weight in man were measurable up to 8 hours post dosing, Figure 7.

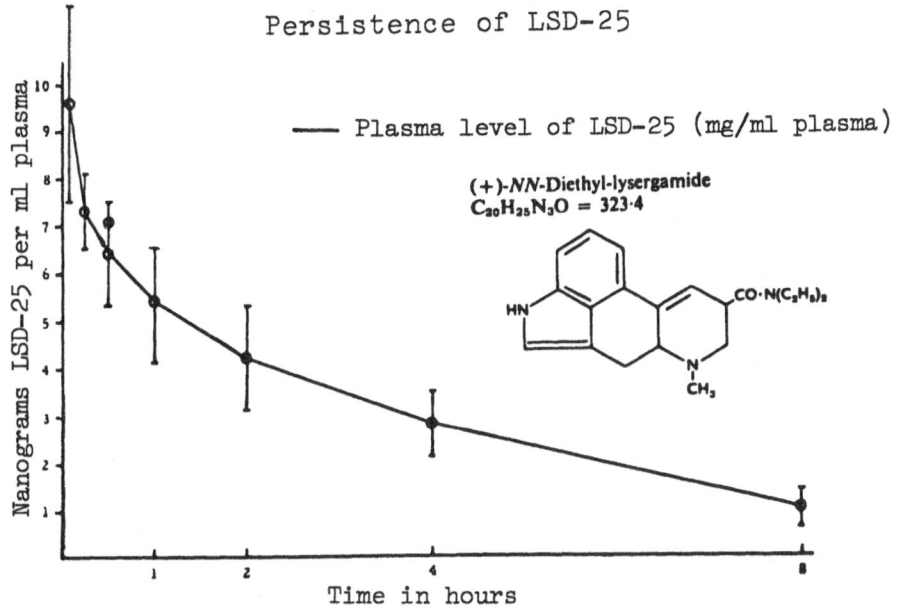

Figure 7. Blood levels of lysergic acid diethylamide (LSD) in man.

    Chlorprothixene (a thioxanthene analog) and
the phenothiazines, such as chlorpromazine and
thioridazine develop intense fluorescence in cold
concentrated H_2SO_4. The fluorophor formed is
probably due to a protonated species formed by
dehydrogenation of the tricyclic ring as shown for
chlorprothixene in Figure 8.

    Chlorprothixene and its sulfoxide metabolite
are extracted from blood at alkaline pH (11-13)
into n-heptane containing 1% isoamyl or isopentyl
alcohol added as a surface deactivating agent.
The sulfoxide can be removed by back extraction
into pH 5.0 buffer, while the parent drug remaining in the organic layer is back extracted into
2 ml of 40% H_2SO_4. The method is, therefore,
specific for the parent compound, and was used in
pharmacokinetic studies in man (28).

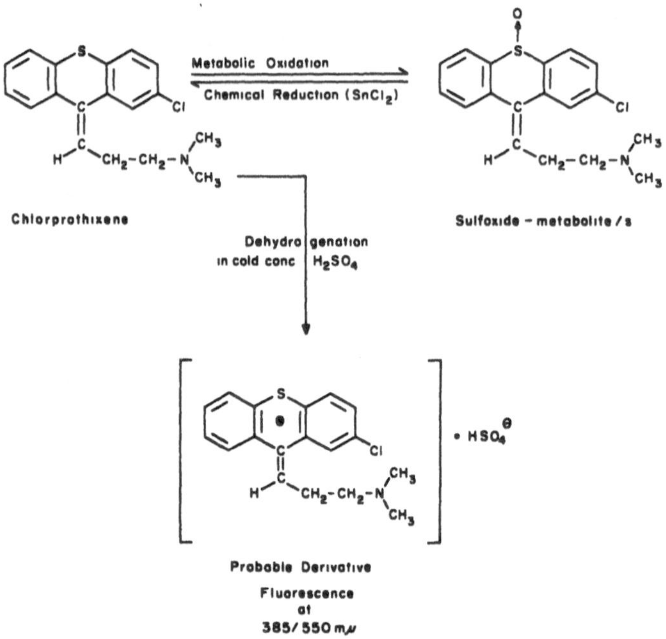

Figure 8.   Chemical reactions of chlorprothixene.

The intense native fluorescence of a number
of polynuclear aromatic or heterocyclic compounds
has been used in their quantitation in biological
fluids.   These include the carbazoles (anti-
inflammatory agents) (29), quinolines (antimalarials) (30), quinazolinones, such as metha-
qualone, a tranquilizer (31), the ergotamine
alkaloids (32), and the amino-acridine analogs
(antimicrobial agents) (33).

B. Analysis using derivatization reactions to produce a fluorophor

Organic (drug) compounds which either have
weak or no intrinsic fluorescence can be deriva-
tized to produce highly fluorescent products which
can be isolated and analyzed (6).   Most of these
derivatization reactions yield products with
increased aromaticity in the molecule resulting
in fluorescent species due to enhanced U.V.

absorption. A few interesting examples of general applicability to drug analysis are listed below:

(i) <u>Oxidation reactions</u> have resulted in a variety of fluorescent derivatives. The phenothiazines and thioxanthenes can be oxidized with $KMnO_4$ in alkaline or acidic media to produce the sulfoxides or sulfones which are fluorescent derivatives (34, 35), Figure 9-A and B.

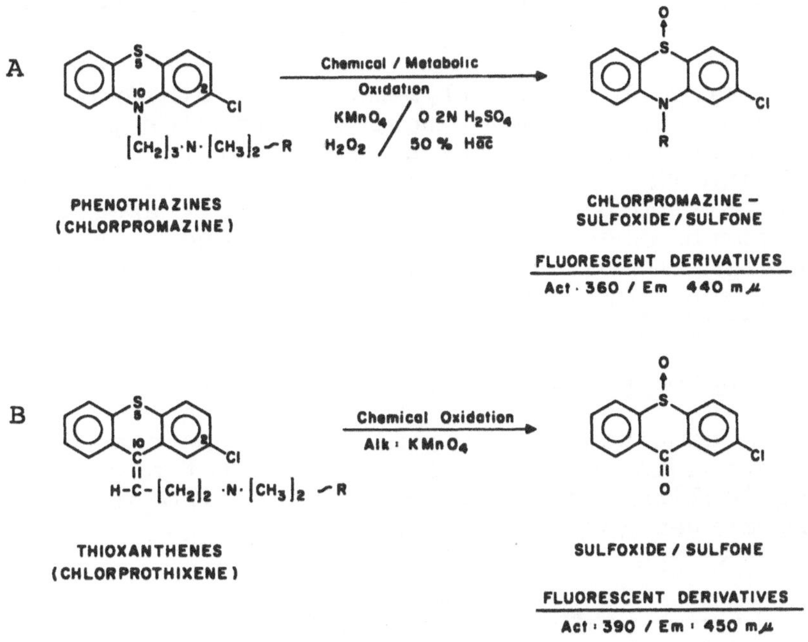

A

Chemical / Metabolic
Oxidation
$KMnO_4$ / $O\ 2N\ H_2SO_4$
H_2O_2 / 50 % H$\overline{ac}$

PHENOTHIAZINES
(CHLORPROMAZINE)

CHLORPROMAZINE –
SULFOXIDE / SULFONE

FLUORESCENT DERIVATIVES
Act · 360 / Em 440 mμ

B

Chemical Oxidation
Alk · $KMnO_4$

THIOXANTHENES
(CHLORPROTHIXENE)

SULFOXIDE / SULFONE

FLUORESCENT DERIVATIVES
Act · 390 / Em · 450 mμ

Figure 9. Fluorescent derivatives of (A) Phenothiazines and (B) Thioxanthenes.

Although morphine exhibits intrinsic fluorescence in dilute acid (0.1 N H_2SO_4), oxidation in base with potassium ferricyanide yields the highly fluorescent derivative pseudomorphine with excitation-emission maxima at 250 and 440 nm respectively (36). This reaction yielded one of more sensitive chemical assays for morphine yet developed, Figure 10.

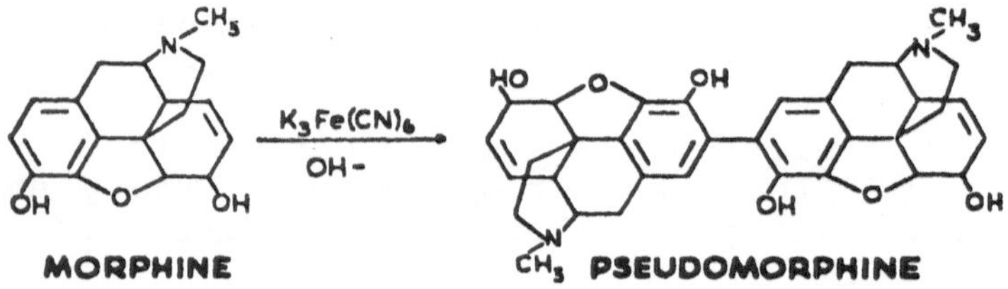
CH3
N
OH
O
OH
K3Fe(CN)6
OH-
HO
O
OH
CH3
N
OH
O
OH
N
CH3
MORPHINE
PSEUDOMORPHINE

Figure 10. Oxidation of morphine in base to yield the highly fluorescent derivative pseudomorphine.

Normorphine, N-allyl-morphine, dihydro morphine, and 6-acetyl-morphine interfere, since they also yield a derivative similar to pseudomorphine which fluoresces. These derivatives can be separated by TLC analysis to impart specificity. The morphinanes, codeine, and its congeners, diacetyl morphine, apomorphine, meperidine, methadone, and dihydromorphinone do not interfere with the assay (36).

(ii) Photochemical oxidation

A combination of chemical and photochemical oxidation in acidic media converts tetrahydroisoquinolines to highly fluorescent isoquinolinium derivatives (37), Figure 11.

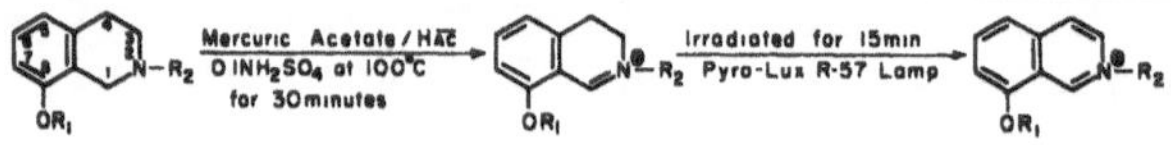
N-R2
OR1
Mercuric Acetate / HAc
0.1N H_2SO_4 at 100°C
for 30 minutes
N⊕-R2
OR1
Irradiated for 15 min
Pyro-Lux R-57 Lamp
N⊕-R2
OR1

R-GROUP		TETRAHYDROISOQUINOLINE —CHEMICAL OXIDATION→ 3,4-DIHYDRODERIVATIVE —PHOTOCHEMICAL OXIDATION→ ISOQUINOLINIUM DERIVATIVE			WAVE LENGTHS (nm) OF	
		PARENT COMPOUND	INTERMEDIATE	FLUOROPHOR	ACTIVATION / EMISSION	
R_1	R_2					
$-CH_3$	$-CH_3$	[II]	[II-A]	[II-B]	370	458
$-CH_3$	-H	[III]	[III-A]	[III-B]	370	458
-H	$-CH_3$	[IV]	[IV-A]	[IV-B]	425	525
-H	-H	[V]	[V-A]	[V-B]	425	525

Figure 11. Chemical and photochemical oxidation of tetrahydroisoquinolines to yield highly fluorescent isoquinolinium derivatives.

The initial chemical oxidation step introduces a double bond in the ring to form the 3,4-dihydro-isoquinolinium derivative which is fluorescent. Photochemical oxidation of this intermediate to the fully aromatized isoquinolinium derivative is achieved by irradiation to U.V. energy from a Pyrolux-R-57 lamp (Luxor Corp., N. Y.). Although the latter derivative has the same excitation-emission maxima as the intermediate, the luminescence intensity, however, is enhanced several fold imparting high sensitivity to the analysis.

Metabolites formed by N-demethylation of the parent compound can be easily distinguished from those formed by O-demethylation which result in phenolic compounds, since their isoquinolinium derivatives have markedly different excitation-emission maxima. This reaction is also applicable as a rapid identification procedure for tetrahydro-isoquinolines following thin layer chromatographic separation. The plate is lightly oversprayed with the mercuric acetate-acetic acid-H_2SO_4 mixture, heated at 100°C in an oven for 10 to 15 minutes and then observed under short and long wave U.V. irradiation. Characteristic blue-green or orange-red fluorescent spots are observed identifying the compounds and their metabolites which may be quantitated either in situ by spectrofluorodensitometry or following elution by spectrofluorometry.

(iii) Chemical hydrolysis, cyclization, and re-arrangement reactions

The 1,4-benzodiazepines and their 2-ones are widely used as tranquilizers, muscle relaxants, sedatives, and hypnotics (38). These compounds can be chemically manipulated in a number of different ways to produce fluorescent derivatives, Figure 12. They can be hydrolyzed in strong acids (6 N HCl) to yield the O-aminobenzophenones which can then be cyclized in base (DMF/K_2CO_3), (especially those containing a fluoro group in the 2'-position of the 5-phenyl ring) to form the highly fluorescent 9-acridanones. Compounds such as medazepam and diazepam fluoresce in cold

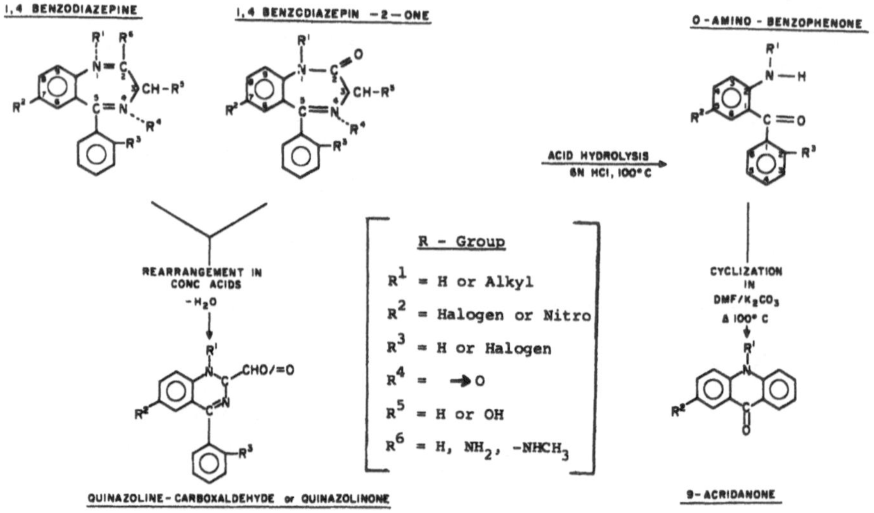

Figure 12. Chemical reactions of 1,4-benzodiaze-pines and their 2-ones to yield fluorescent derivatives.

concentrated acids probably due to protonation of the 1,4-benzodiazepine ring (39), while the N-desalkyl-3-hydroxy analogs, oxazepam, and loraze-pam undergo dehydration in concentrated acids, such as perchloric, sulfuric, or orthophosphoric acids. The dehydration reaction results in a reduction of the 7-membered ring to the highly fluorescent 6-membered quinazolinone or quinazo-line-carboxaldehyde derivatives which can be analyzed with submicrogram sensitivity (40). This is a very useful identification tool in that a thin layer chromatoplate containing these com-pounds can be oversprayed with either acid and examined under short and long wave U.V. to observe the characteristic blue-green to yellow fluorescence of these compounds.

Chlordiazepoxide, the first benzodiazepine drug to be marketed, is analyzed fluorometrically by mild acid hydrolysis to demoxepam (also a metabolite) which is then converted photochemi-cally in base (0.1 N NaOH) to the highly fluor-escent quinazolinone derivative. The major

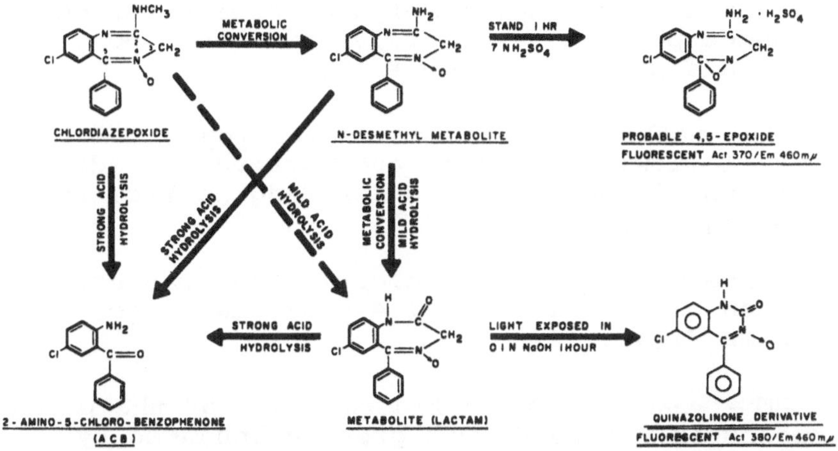

Figure 13. Chemical reactions of chlordiazepoxide and its major metabolites.

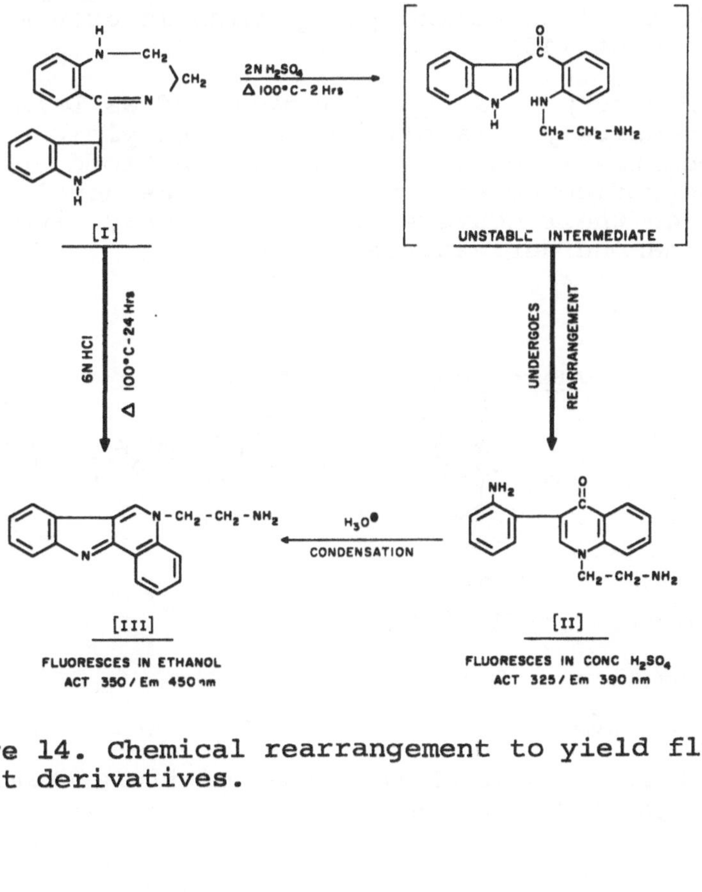

Figure 14. Chemical rearrangement to yield fluorescent derivatives.

metabolites viz., N-desmethyl chlordiazepoxide
and demoxepam are selectively extracted and
analyzed separately, Figure 13.

The method was used in monitoring blood levels
in man following the oral and intravenous administration of therapeutic doses ranging from 25
to 100 mg (41).

A very interesting molecular rearrangement
was seen with an indolyl-1,4-benzodiazepine,
shown in Figure 14.

The weak intrinsic fluorescence of the parent
compound in 2 N H_2SO_4 was greatly enhanced by
heating at 100°C for 2 hours, whereby the highly
fluorescent aminophenylquinolone [II] derivative
was formed. The latter [II] can also be cyclized
to the indoloquinoline [III] which is even more
fluorescent (42).

Flurazepam (a hypnotic marketed as Dalmane[R])
is extensively metabolized in man to yield several
metabolites which are measurable in blood and urine.
A spectrofluorometric method (43) was developed
based on their conversion to the highly fluorescent
9-acridanone derivatives, Figure 15.

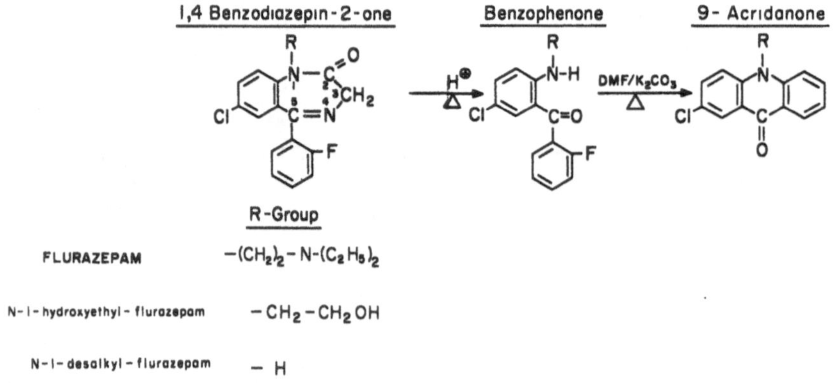

Figure 15. Chemical reactions of flurazepam and
its major metabolites.

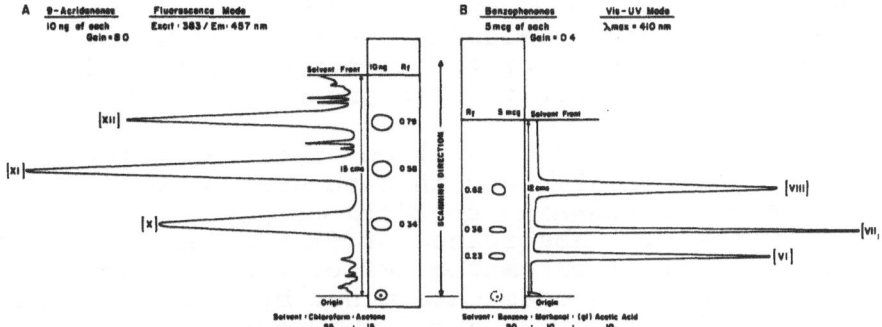

Figure 16.   Chromatograms of the 9-acridanones and of the benzophenones.

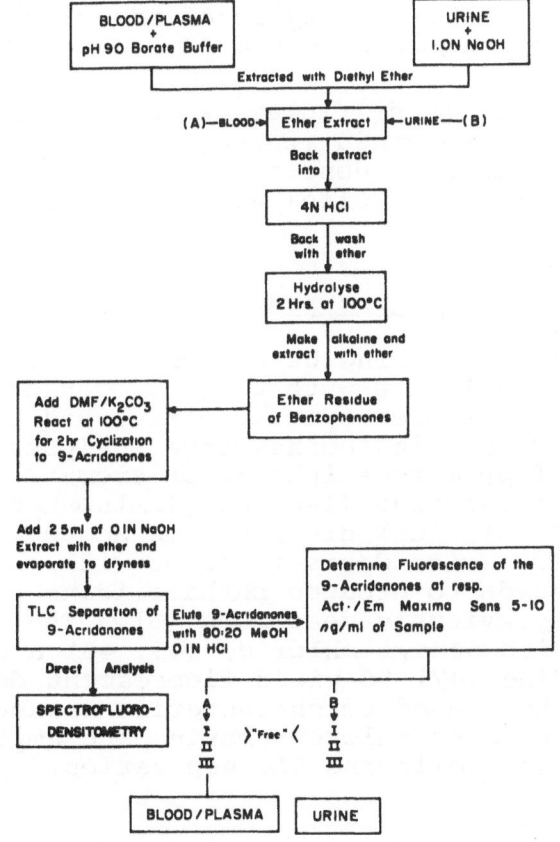

Scheme 1.   Flow diagram of procedure for the determination of flurazepam and its major metabolites.

The three main components in blood are the
hydroxyethyl and N-desalkyl metabolites along
with trace amounts of the parent compound which
yield their respective acridanones.

Following TLC separation of the 9-acridanones,
they can be quantitated in situ using a thin layer
chromatogram scanning densitometer operated in
either the U.V. absorbance or fluorescence emission
modes. Typical chromatograms of the 9-acridanones
and the benzophenones determined by spectrofluoro/
photodensitometry are shown in Figure 16.

The flow diagram of the analytical procedure
is shown in Scheme 1, and the sensitivity of the
fluorodensitometric assay is of the order of 1 to
2 ng of compound/ml of blood.

Blood levels of flurazepam and its major
metabolites determined in two subjects following
a single 30 mg oral dose are shown in Figure 17
and demonstrate the clinical utility of the assay
(44).

(iv) Chemical coupling reactions

Chemical coupling or condensation of a nonfluorescent molecule with a "fluorogenic" reagent
to produce a fluorescent derivative has been
extensively used in luminescence analysis. The
coupling of primary aliphatic or aromatic amines
with dansyl chloride (1-dimethylaminonaphthalene-
5-sulfonyl chloride) dissolved in acetone to form
dansyl-sulfonamide derivatives or with ortho-
phthalaldehyde to produce phthalimidine derivatives
have been previously described (6). The coupling
reaction with dansyl chloride (45, 46) and with
cyanoacridine (47) to yield fluorescent derivatives has been used to characterize metabolites
of drugs, such as chloropromazine and amytripty-
line in urine following TLC separation.

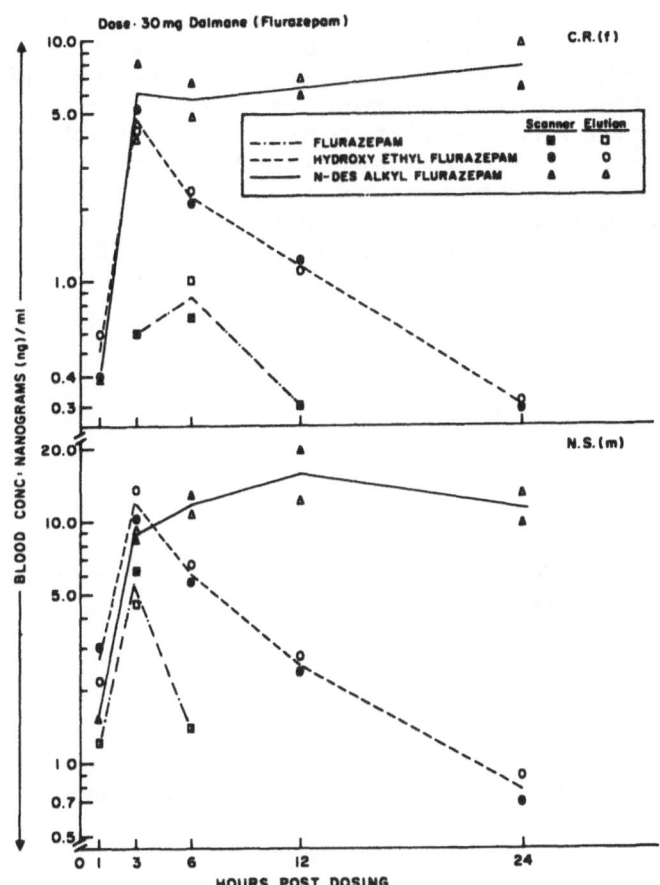

Figure 17.   Blood levels of flurazepam and its major metabolites in man following a single 30 mg oral dose.

The main disadvantage of these reactions is
that excess reagent which is intrinsically fluorescent must be separated from the derivative produced prior to analysis.

A recently developed "fluorogenic" reagent
called fluorescamine (Fluram[R]) reacts specifically with primary aliphatic or aromatic amines
to produce highly fluorescent derivatives which
can be quantitated with picomole to nanomole
sensitivity. The reagent itself is nonfluorescent,
and the excess reagent is readily hydrolyzed to
nonfluorescent products. The chemical reactions
of fluorescamine are shown in Figure 18.

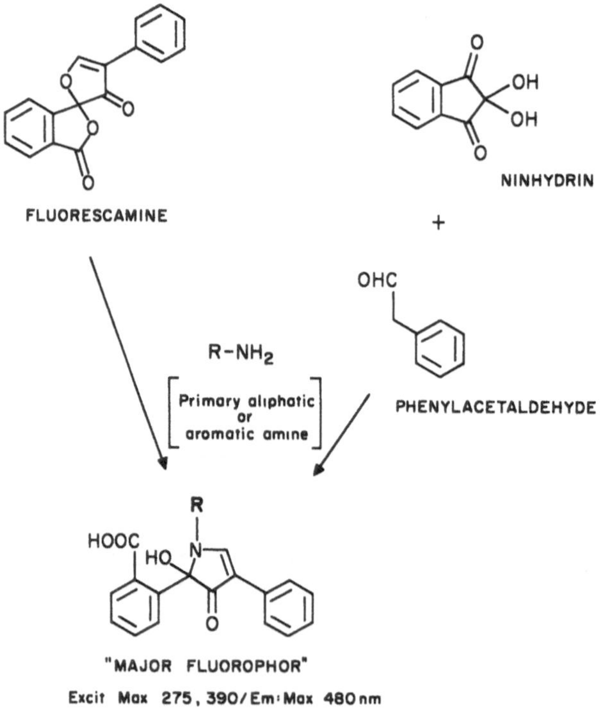

Figure 18. Chemical reactions of fluorescamine.

The chemical structure of the fluorophor formed with fluorescamine is identical to that formed by the reaction of ninhydrin with amino acids or peptides in the presence of phenyl acetaldehyde (48). The nonspecific nature of the derivative formed yields similar excitation-emission spectra, hence a chromatographic separation step is required for specificity of analysis. The sensitivity of quantitation may be enhanced by selective extraction of the derivative into ethyl acetate by adjusting the pH of the reaction medium to 5.0. A variety of pharmaceuticals can be determined by this technique (49).

Fluorescamine has revolutionized amino acid and peptide analysis (50) in that the reaction is ideally suited to monitoring the effluent of cation or anion exchange resin columns used in their greatly accelerated analysis by high pressure liquid-liquid chromatography (HPLC).

The use of low dead volume capillary flow cells and fluorometric detectors of high sensitivity has enabled accurate quantitation of amino acids at picomole (ng) to nanomole (μg) sensitivity limits. Drug molecules, such as the catecholamines (dopamine), amphetamines, sulfonamides, and other aromatic amines which are polar and difficult to derivatize for GLC analysis can be readily analyzed by HPLC using fluorescamine as the fluorogenic reagent.

V. Application of phosphorescence analysis to drugs

Phosphorescence analysis of drugs in biological media has not been widely used until recently, partly because of the fact that it is a somewhat more complicated technique than fluorescence and partly due to limitations in instrumentation. The technique, however, is a most useful one and complements fluorometric analysis in terms of sensitivity and specificity which is further enhanced by the ability to measure decay lifetimes (51, 52). The work of Winefordner and

his associates is particularly noteworthy in
their efforts to stimulate the use of phosphorescence in drug analysis (53).

The endogenous phosphorescent products extracted into diethyl ether from blood and urine
at various pH values which contribute to the
background signal are shown in Figure 19.

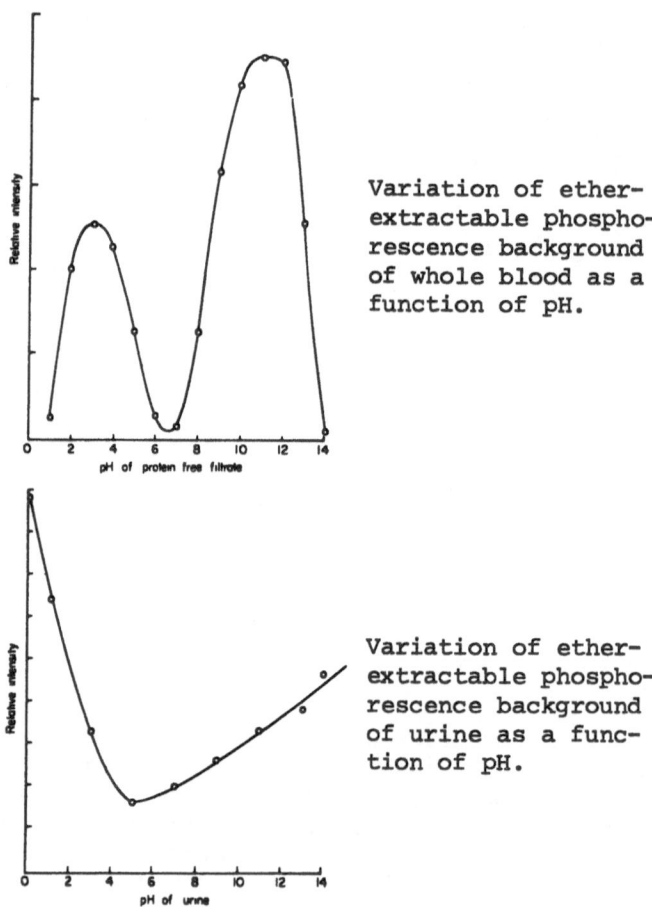

Variation of ether-
extractable phospho-
rescence background
of whole blood as a
function of pH.

Variation of ether-
extractable phospho-
rescence background
of urine as a func-
tion of pH.

Figure 19.   Variation of phosphorescent background with pH due to extractable endogenous
impurities.

The "cleanest" blood extracts appear to be obtained at pH's 1, 7, and 14; hence a wide spectrum of drugs which are quantitatively extracted at these pH's can potentially be analyzed by phosphorescence. A simple technique is to obtain a protein free filtrate of blood using absolute ethanol in which most drugs are soluble and analyze the filtrate directly as a "glass" after drying the solution with anhydrous Na_2SO_4 or $MgSO_4$. This step removes moisture which would otherwise cause vitrification or "cracking" of the glass producing significant scattering effects (53).

The analysis of urine on the other hand presents more problems, since significant blanks are obtained throughout the pH range with a minimum at around pH 5.0. However, TLC analysis can be used to "clean-up" the extract prior to elution of the drug into either absolute ethanol or EPA [diethyl ether:isopentane:ethanol (5:5:2)] which are the solvents most widely used in phosphorescence analysis at 77°K.

Fluorescence and phosphorescence spectra of selected tetrahydrocarbazoles, carbazoles, 1,4-benzodiazepines, and analytically useful derivatives of the 1,4-benzodiazepines were determined at 77°K using the modified Farrand Mark I Spectrofluorometer.

(A) Indole Analogs: Compounds containing the indole nucleus have well defined intrinsic phosphorescent properties. The spectra of indole-3-acetic acid and of indomethacin, a substituted indole used as an anti-inflammatory and analgesic agent are shown in Figure 20.

A simple extraction procedure for indomethacin which includes a TLC separation step to ensure specificity is shown in Scheme 2. The sensitivity limits of the assay are of the order of 0.5 μg/ml of blood.

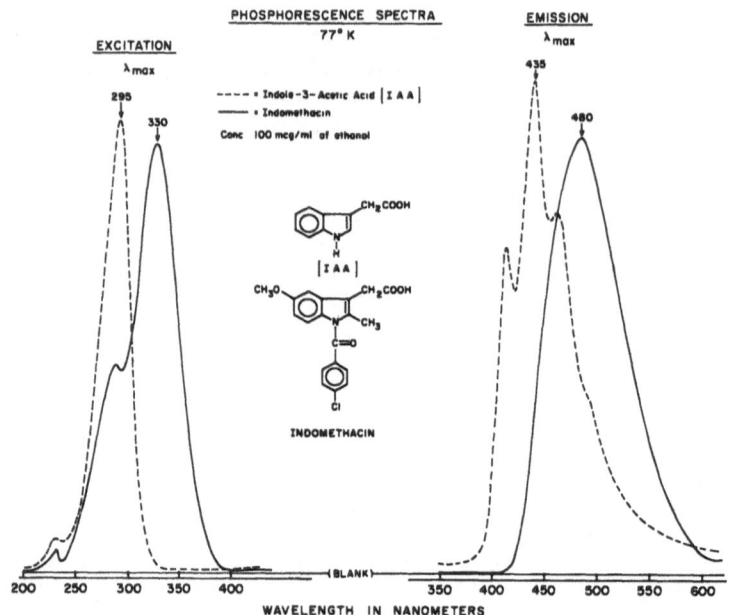

Figure 20. Phosphorescence spectra of indole-3-acetic acid and indomethacin.

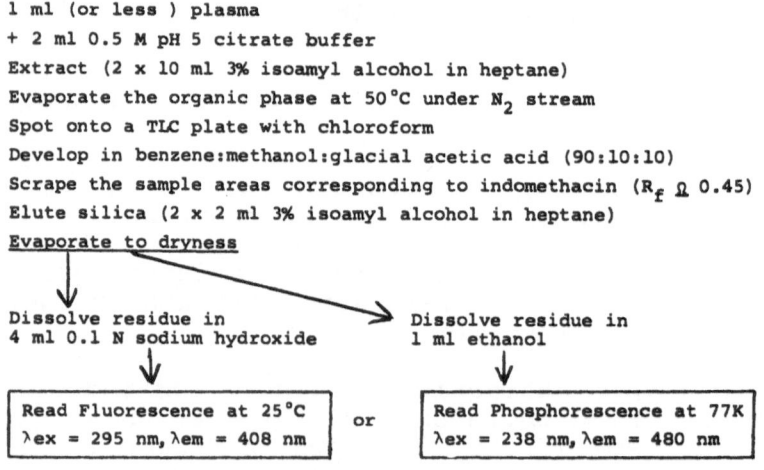

Scheme 2. Flow diagram of the analytical procedure for indomethacin.

(B) <u>Carbazoles:</u> The substituted tetrahydrocarbazoles are especially interesting as they have
little intrinsic fluorescence at room temperature,
whereas they exhibit intense fluorescence and
phosphorescence at cryogenic temperatures (77°K).
The luminescence spectra of 6-chloro-1,2,3,4-
tetrahydrocarbazole-2-carboxylic acid, an anti-
inflammatory agent, are shown in Figure 21.

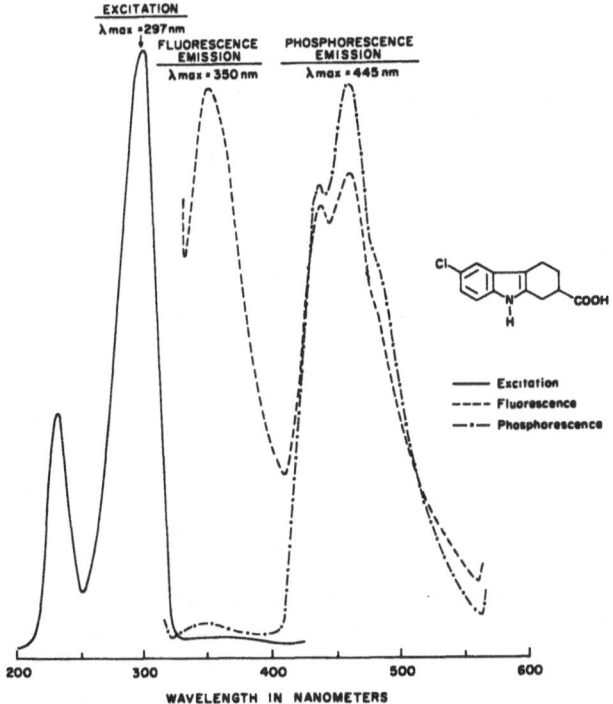

Figure 21. Fluorescence and phosphorescence
spectra of 6-chloro-1,2,3,4-tetrahydrocarbazole-
2-carboxylic acid (1.0 µg/ml) in ethanol at 77°K.

The sensitivity of the phosphorescence emission
is sufficient for the determination of 0.5 to
1.0 µg of compound/ml of blood, and was sufficient
for use in clinical studies (54).

The phosphorescence spectra of another anti-inflammatory agent, 6-chloro-α-methylcarbazole-2-acetic acid, are shown in Figure 22.

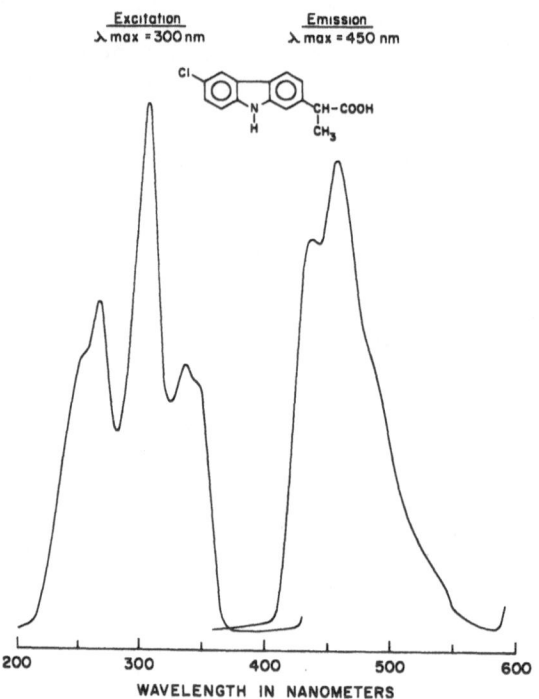

Figure 22.   Phosphorescence spectra of (d,l)-6-chloro-α-methylcarbazole-2-acetic acid (1.0 µg/ml) in ethanol at 77°K.

Although there are no obvious changes seen in the "fine" structure of the emission spectrum compared with that of the tetrahydrocarbazole, Figure 21, significant changes in the "fine" structure of the excitation spectrum as multiple peaks are clearly evident due to the increased aromaticity in the molecule (52). Thus phosphorescence spectra can yield valuable information relative to structural changes and the effects of substituent groups in the molecule.

The linear dynamic range curves of phosphorescence intensity vs. concentration of the 6-chlorocarbazole are shown in Figure 23 for each of three instruments examined.

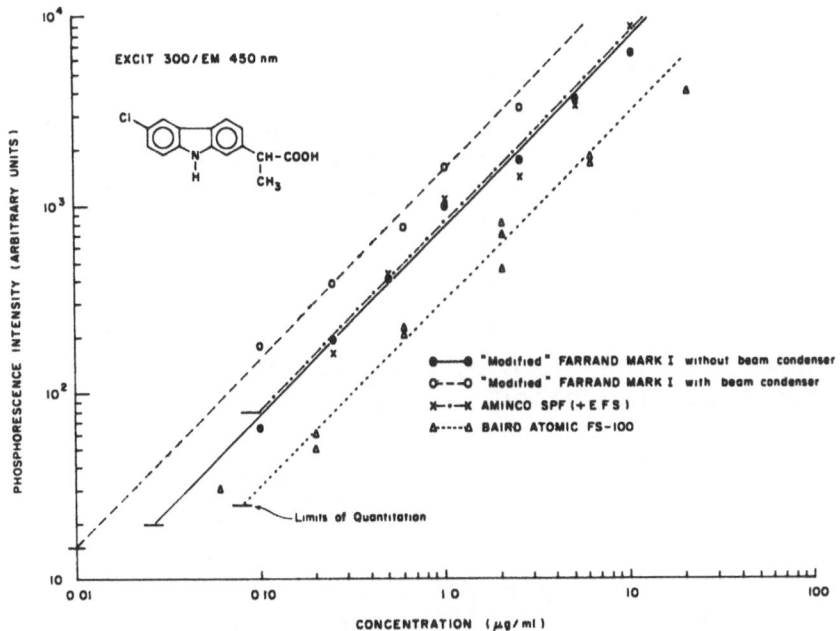

Figure 23. Phosphorescence calibration of (d,l)-6-chloro-α-methylcarbazole-2-acetic acid in EPA at 77°K.

The useful range of phosphorescence measurement covers at least three orders of magnitude and also shows the modified Farrand to be equal to or better than the Aminco SPF and Baird Atomic spectrofluorometers in linearity and sensitivity.

A typical simplified procedure for the extraction of carbazoles from blood and analysis by fluorescence and phosphorescence is shown in Scheme 3.

The sensitivity limit for either method is of the order of 0.2 to 0.4 μg of compound/ml of

Analytical Procedure for (d,l)-6-chloro-α-
methylcarbazole-2-acetic acid

1 ml blood or plasma
+ 5 ml 1.0 M pH 6.8 buffer
Extract (2 x 10 ml diethyl ether), combine extracts
Evaporate (30-35° under N$_2$ stream)
Spot on TLC plate with chloroform
Develop in chloroform:diethylamine (90:10), 15 cm ascending
Air-dry 5 min.
Develop in chloroform:ethanol:formic acid (90:10:5), 15 cm ascending
Scrape sample area corresponding to authentic standards (R$_f$ ≈ 0.35)
Elute silica (1 x 5 ml 1% acetic acid in ethanol)

| Read Fluorescence at 25°C
λex = 300 nm, λem = 370 nm | or | Read Phosphorescence at 77K
λex = 302 nm, λem = 455 nm |

Scheme 3. Flow diagram for the extraction of carbazoles from blood.

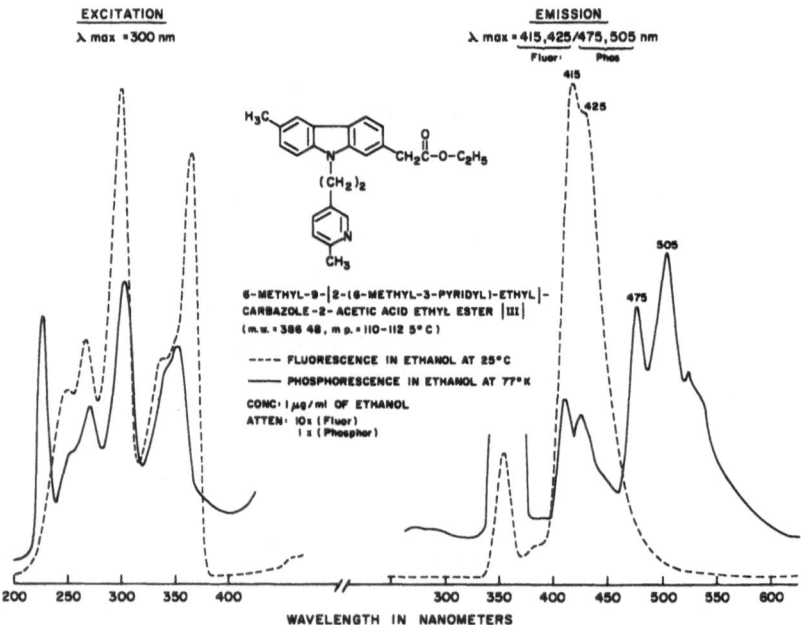

Figure 24. The influence of substituent groups on the luminescence spectra of a carbazole.

blood. The TLC separation step ensures both
sample "clean-up" and assay specificity in resolv-
ing the parent compound from any metabolites
present and from endogenous interferences.

The influence of substituent groups on the
luminescence of the carbazole nucleus is shown in
Figure 24 for 6-methyl-9-[2-(6-methyl-3-pyridyl)-
ethyl]-carbazole-2-acetic acid ethyl ester.
The aromaticity of the carbazole nucleus is
greatly increased due to the electron denoting
substituent group attached to the carbazole ring
resulting in greatly increased "fine structure"
in both the excitation and emission spectra of the
compound.

(C) <u>1,4-Benzodiazepines</u>: Luminescence data on the
1,4-benzodiazepines were obtained in 1% (36 N)

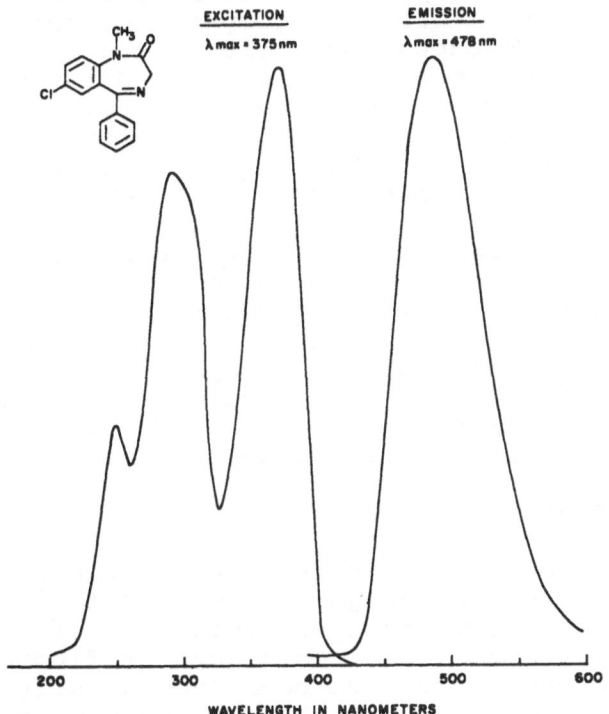

Figure 25. Phosphorescence spectra of diazepam
(1 µg/ml) in 1% H_2SO_4 (36N) in ethanol at 77°K.

H_2SO_4 in ethanol, because EPA and ethanol gave
lower luminescence intensity than did acidified
ethanol (16).

The 2,3-dihydro compounds, medazepam and its
N-desmethyl metabolite gave similar spectra but
widely differing intensities, whereas compounds
having the N_4-oxide, such as chlordiazepoxide
showed little luminescence.

Diazepam and N-desmethyl diazepam showed the
highest phosphorescence intensity and their
spectra are characteristic of most of the 1,4-
benzodiazepin-2-ones (Figure 25).

The 7-nitro and 7-amino benzodiazepin-2-
ones showed distinctive shifts in their maxima.
The luminescence observed as "phosphorescence"
appears to be delayed fluorescence, since no
change in maxima occurred compared to those recorded for "instantaneous" fluorescence at 77K.
The phosphorescence of the 1,4-benzodiazepin-2-
ones is relatively shortlived; diazepam, medazepam,
bromazepam, and flurazepam showed lifetimes
shorter than 0.2 seconds, while clonazepam showed
a lifetime of between 0.4 and 0.6 seconds (16).

(D) Phosphorescent derivatives. No specific
chemical reactions that produce phosphorescent
derivatives only have been reported, although
most of the reactions used to produce fluorescent
may also be applicable to phosphorescence. Reactions that produce either chemi, bio, thermo,
or electrogenerated luminescence may be considered
under this category of phosphorescent derivatives.

(VI) Epilogue

Luminescence analysis is particularly useful
in the analysis of drugs in biological fluids
where sensitivity and specificity are of para-
mount importance. The techniques of fluorescence
and phosphorescence have been successfully used
in the determination of a number of drug classes,
such as the antimalarials (55, 56), cannabinols
(57), hallucinogens (58), barbiturates (59), anti-

histamines (60), sulfonamines (61), and vitamins (62). Recent studies have extended its applicability to other classes of drugs, such as the tetrahydrocarbazoles, carbazoles, 1,4-benzodiazepines, and to the analytically useful derivatives of the 1,4-benzodiazepines, viz., the benzophenones, 9-acridanones, and quinazolines (16). The application of luminescence in clinical chemistry has also been recently reviewed (63).

Instrument design modifications have allowed the use of several recent improvements in phosphorimetric techniques, such as the use of the NMR sample tube spinning apparatus for improved precision through better statistical signal averaging, and the use of open ended capillary cells which extend the use of phosphorimetry to solvents which form either cracked glasses or snowed matrices and to analysis in aqueous solutions. While low temperature luminescence spectroscopy is not as simple a technique as room temperature fluorescence, it can be used on a routine basis to good advantage in special applications which merit its use (64).

Advances in instrumentation have effected a wider application of luminescence methods to drug analysis. Among these are the use of lasers as excitation sources in specially designed spectrofluorometers (65-67), the use of U.V. or fluorescence detectors in high performance liquid chromatography (HPLC), and the use of spectrofluorodensitometry for "in situ" quantitation of drugs following thin layer chromatographic separation (TLC).

Luminescence techniques, because of their inherent high sensitivity and specificity, are eminently suitable for the analysis of drugs at nanogram to picogram concentrations. These detection limits may soon be extended to the femtogram (10^{-15} g) range using laser excited spectrofluorometry.

ACKNOWLEDGMENT

The author thanks the following for their permission to reproduce the figures below which are copyright materials.

Figure 1 - Analytica Chimica Acta, Elsevier
           Scientific Publishing Co., Amsterdam,
           Netherlands.

Figure 2 - Farrand Optical Co., Valhalla, N. Y.

Figure 3, 19 - Academic Press, N. Y., N. Y.

Figures 4, 12, 14, 18, 23, 24 - Analytical Chemis-
           try, American Chemical Society, Wash-
           ington, D.C.

Figures 5, 8, 13 - J. Forensic Science - Callaghan
           & Co., Mundelein, Illinois.

Figures 11, 15, 16, 17 - J. Pharmaceutical
           Sciences, American Pharmaceutical
           Association, Washington, D.C.

Figures 20, 21 - Schemes 2, 3 - J. Chromato-
           graphic Sciences, Preston Technical
           Abstracts Co., Niles, Illinois.

Figure 7 - Clinical Pharmacology and Therapeutics.
           The C. V. Mosby Co., St. Louis,
           Missouri.

Figure 10   J. Pharmacology and Experimental Thera-
           peutics, The Williams and Wilkins Co.,
           Baltimore, Maryland.

REFERENCES

1. J.A.F. de Silva, "Current Concepts in the Pharmaceutical Sciences", Vol. I - Biopharmaceutics, (Ed., J. Swarbrick), Lea and Febiger, Philadelphia, pp. 203-264, 1970.

2. C. A. Parker, "Photoluminescence of Solutions with Applications to Photochemistry and Analytical Chemistry", American Elsevier Publishing Co., New York, 1968.

3. R. A. Becker, "Theory and Interpretation of Fluorescence and Phosphorescence", Wiley - Interscience Publishers, New York, 1969.

4. G. H. Schenk, "Absorption of Light and Ultra Violet Radiation", Fluorescence and Phosphorescence Emission, Allyn and Bacon, Inc., Boston, 1973.

5. D. MacDougall, Residue Reviews, $\underline{5}$, 119, 1966.

6. J.A.F. de Silva and L. D'Arconte, J. Forensic Sci., $\underline{14}$, 184, 1969.

7. C.N.R. Rao, Ultraviolet and Visible Spectroscopy-Chemical Applications, Butterworth & Co., London, pp. 39-60, 1961.

8. E. Sawicki, Talanta, $\underline{16}$, 1231, 1969.

9. J. Bartos and M. Pesez, Talanta, $\underline{19}$, 93, 1972.

0. J.A.F. de Silva, N. Strojny, and N. Munno, Anal. Chim. Acta, $\underline{66}$, 23, 1973.

1. E. Sawicki and J. D. Pfaff, Anal. Chim. Acta, $\underline{32}$, 521, 1965.

2. J.S.T. Chou and B. M. Lawrence, J. Chromatog., $\underline{27}$, 279 (1967).

13. H. P. Raaen and L. J. Crist, J. Chromatog.,
    <u>39</u>, 515, 1969.

14. B. L. van Duuren and T. L. Chan., "Fluores-
    cence Spectrometry" in J. D. Winefordner
    (Ed) Spectrochemical Methods of Analysis,
    Chemistry and Instrumentation, J. Wiley &
    Sons, pp. 387-450, 1971.

15. W. J. McCarthy, "Phosphorescence Spectro-
    metry", ibid, J. Wiley & Sons, pp. 451-492,
    1971.

16. J.A.F. de Silva, N. Strojny, and K. Stika,
    Anal. Chem., <u>48</u>, 144, 1976.

17. J. D. Winefordner, S. G. Schulman, and T.
    C. O'Haver, "Luminescence Spectrometry in
    Analytical Chemistry", Wiley Interscience
    Publishers, New York, pp. 269-314, 1972.

18. C. A. Parker and W. T. Rees, Analyst, <u>87</u>,
    83, 1962.

19. C. A. Parker and C. A. Hatchard, Analyst,
    <u>87</u>, 664, 1962.

20. T. J. Porro, R. E. Anacreon, P. S. Flandreau,
    and I. S. Fagerson, J. Off. Anal. Chem.
    (JOAC), <u>56</u>, 607, 1973.

21. J. W. Eastman, Photochem; Photobiol., <u>6</u>, 55,
    1967.

22. G. Schill in "Ion Exchange and Solvent Ex-
    traction", J. A. Marinsky and Y. Marcus (Eds),
    Marcel Dekker Inc., New York, Vol. 6, pp. 1-
    57, 1974.

23. R. E. Huettemann, M. L. Cotter, C. J. Shaw,
    C. A. Janicki, H. R. Almond, E. S. Moyer,
    A. P. Shroff, and F. Vestano, Pharmaceuticals
    and Related Drugs, Anal. Chem., <u>47</u> (5), 233R-
    289R, 1975.

24. S. Udenfriend, Fluorescence Assay in Biology and Medicine, Vols. I and II, Academic Press, New York, pp. 400-443, 1962, and pp. 503-538, 1969.

25. M. Chirigos and S. Udenfriend, J. Lab. Clin. Med., 54, 769, 1959.

26. S. Udenfriend, D. E. Duggan, B. M. Vasta, and B. B. Brodie, J. Pharmacol. Exptl. Therap., 120, 26, 1957.

27. G. K. Aghajanian and O.H.L. Bing. Clin. Pharm. Therap., 5, 611, 1966.

28. J. Raaflaub, Experentia, 31, 557, 1975.

29. J.A.F. de Silva, N. Strojny, and M. A. Brooks, Anal. Chim. Acta, 73, 283, 1974.

30. S. G. Schulman and J. F. Young, Anal. Chim. Acta, 70, 229, 1974.

31. S. G. Schulman, J. M. Rutledge, and G. Torosian, Anal. Chim. Acta, 68, 455, 1974.

32. W. D. Hooper, J. M. Sutherland, M. J. Eadie, and J. H. Tyrer, Anal. Chim. Acta, 69, 11, 1974.

33. D. V. Naik and S. G. Schulman, Anal. Chim. Acta, 80, 67, 1975.

34. T. J. Mellinger and C. E. Keeler, Anal. Chem., 35, 554, 1962, and Anal. Chem. 36, 1840, 1964.

35. J. B. Ragland and V. J. Kinross-Wright, Anal. Chem., 36, 1357, 1964.

36. H. J. Kupferberg, A. Burkhalter, and E. L. Way, J. Pharmacol. Expt. Therap., 145, 247, 1964.

37. J.A.F. de Silva, N. Strojny, and N. Munno, J. Pharm. Sci., 62, 1066, 1973.

38. D. J. Greenblatt and R. I. Shader, "Benzo-
    diazepines in Clinical Practice, Raven
    Press, N. Y., 1974.

39. D. D. Maness and G. J. Yakatan, J. Pharm.
    Sci., 64, 651, 1975.

40. G. Caille, J. Braun, D. Gravel, and R. Plourde,
    Canad. J. Pharm. Sci., 8, 42, 1973.

41. M. A. Schwartz, E. Postma, and Z. Gaut, J.
    Pharm. Sci., 60, 1500, 1971.

42. J.A.F. de Silva, N. Munno, and N. Strojny,
    Anal. Chem., 45, 665, 1973.

43. J.A.F. de Silva and N. Strojny, J. Pharm. Sci.,
    60, 1303, 1971.

44. J.A.F. de Silva, I. Bekersky, and C. V.
    Puglisi, J. Pharm. Sci., 63, 1837, 1974.

45. P. N. Kaul, M. W. Conway, M. L. Clark, and
    J. Huffine, J. Pharm. Sci., 59, 1745, 1970.

46. I. S. Forrest, S. D. Rose, L. G. Brookes, B.
    Halpern, V. A. Bacon, and I. A. Silberg,
    Agressologie (Paris), 11, 127, 1970.

47. J. E. Sinsheimer, D. D. Hong, J. T. Stewart,
    M. L. Fink, and J. H. Burkhalter, J. Pharm.
    Sci., 60, 141, 1971.

48. M. Weigele, J. F. Blout, J. P. Tengi, R. C.
    Czaijkowski, S. L. De Bernado, and W. Leim-
    gruber, J. Amer. Chem. Soc., 94, 4052 and
    5927, 1972.

49. J.A.F. de Silva and N. Strojny, Anal. Chem.,
    47, 714, 1975.

50. K. Samejima, W. Dairman, J. Stone, and S.
    Udenfriend, Anal. Biochem., 42, 227-247,
    1971.

51. J. D. Winefordner, W. J. McCarthy, and P. A. St. John, "Phosphorimetry as an Analytical Approach in Biochemistry" in D. Glick (Ed.) Methods of Biochemical Analysis, Vol. XV, Interscience Publisher, New York, pp. 369-483, 1967.

52. M. Zander, Phosphorimetry - The Application of Phosphorescence to the Analysis of Organic Compounds, Academic Press, New York, 1968.

53. J. D. Winefordner, P. A. St. John, and W. J. McCarthy, "Phosphorimetry as a Means of Chemical Analysis" in Fluorescence Assay in Biology and Medicine - S. Udenfriend, Vol. II, pp. 42-123, Academic Press, New York, 1969.

54. N. Strojny and J.A.F. de Silva, J. Chromatog. Sci., 13, 583, 1975.

55. S. G. Schulman and L. B. Sanders, Anal. Chim. Acta, 56, 83, 1971.

56. S. G. Schulman, K. Abate, P. J. Kovi, A. C. Capomacchia, and D. Jackman, Anal. Chim. Acta, 65, 59, 1973.

57. A. Bowd, P. Byrom, J. B. Hudson, and J. H. Turnbull, Talanta, 18, 697, 1971.

58. D. M. Fabrick and J. D. Winefordner, Talanta, 20, 1220, 1973.

59. L. A. Gifford, W. P. Hayes, L. A. King, J. N. Miller, D. T. Burns, and J. W. Bridges, Anal. Chem., 46, 94, 1974.

60. D. R. Wirz, D. L. Wilson, and G. H. Schenk, Anal. Chem., 46, 896, 1974.

61. J. W. Bridges, L. A. Gifford, W. P. Hayes, J. N. Miller, and D. Thorburn Burns, Anal. Chem. 46, 1010, 1974.

62. J. J. Aaron and J. D. Winefordner, Talanta,
    19, 21, 1974.

63. C. M. O'Donnell and J. D. Winefordner, Clin.
    Chem., 21, 285, 1975.

64. J. J. Aaron and J. D. Winefordner, Talanta,
    22, 707, 1975.

65. J. P. Webb, Anal. Chem., 44, 30A, 1972

66. J. R. Allkins, Anal. Chem., 47, 752A, 1975.

67. D. C. Harrington and H. V. Malmstadt, Anal.
    Chem., 47, 271, 1975.